AF845899

62 Structure and Bonding

Clusters

With Contributions by
F. A. Cotton G. Gliemann G. Schmid
R. A. Walton H. Yersin

With 94 Figures and 18 Tables

Springer-Verlag
Berlin Heidelberg GmbH

ISBN 978-3-662-15223-2 ISBN 978-3-540-39647-5 (eBook)
DOI 10.1007/978-3-540-39647-5

Library of Congress Catalog Card Number 67-11280

Originally published by Springer-Verlag Berlin Heidelberg New York Tokyo in 1985
Softcover reprint of the hardcover 1st edition 1985

2152/3140-543210

Table of Contents

Metal-Metal Multiple Bonds in Dinuclear Clusters

F. A. Cotton[1] and R. A. Walton[2]

1 Department of Chemistry, Texas A & M University, College Station, Texas 77843, USA
2 Department of Chemistry, Purdue University, West Lafayette, Indiana 47907, USA

Just over 20 years have elapsed since the first recognition of the existence of compounds that contain metal-metal multiple bonds. The present article surveys many of the important developments that have occurred in this field since the publication in 1982 of the text *"Multiple Bonds Between Metal Atoms"* by the present authors. Detailed updates are provided of the chemistry surrounding ditungsten, ditechnetium, and dirhenium complexes that possess quadruple bonds ($\sigma^2\pi^4\delta^2$ electronic configuration) and electron-rich triple bonds ($\sigma^2\pi^4\delta^2\delta^{*2}$ electronic configuration). The rapidly developing chemistry of multiply bonded diruthenium and diosmium complexes that possess the M_2^{4+}, M_2^{5+}, and M_2^{6+} cores, especially those containing ligand bridges, is also surveyed. Additional topics that have been covered are the following: the reactivity of dimolybdenum(III) and ditungsten(III) hexaalkoxides; the chemistry of singly bonded diplatinum(III) complexes; experimental and theoretical aspects of the bonding in $(Cr{\equiv}Cr)^{4+}$ complexes; the diatomics Mo_2, Cr_2, V_2 and Ru_2; and the photoelectron spectroscopy, optical activity, and M–M bond length/bond order and charge correlations that characterize compounds that contain metal-metal multiple bonds.

Structure and Bonding 62

1 Introduction

Our book *Multiple Bonds Between Metal Atoms* was published in March of 1982[1], but systematic coverage of the literature ceased during mid 1981. Since that time the field has continued to expand rapidly. Over the approximately three-year period in question, approximately 350 new research papers have appeared, bringing the total number of publications in the field to well over 1000.

The purpose of this review is to summarize important developments that have occurred since we finished writing our book, which we shall henceforth refer to, for brevity, as MBBMA. In view of the enormous amount of new literature – about half again as many papers as we covered in the book itself – it is obvious that this can only be done selectively. Accordingly, we have chosen a few areas where comprehensive reviewing seemed appropriate and treated them in detail, while in a final section we present an overview of novel discoveries from other parts of the field.

It may be noted that over the past several years reviews covering larger or smaller parts of the subject of multiple bonds between metal atoms have appeared. See, for examples, Refs. 2–8.

2 Detailed Update of Ditechnetium and Dirhenium Chemistry

Ditechnetium Compounds. While the development of ditechnetium chemistry has proceeded relatively slowly over the last few years there have, nonetheless, been some noteworthy advances. Until 1980, all attempts to isolate and characterize the octachloroditechnetate(III) anion had failed, an important reason being the surprising stability of the odd electron $[Tc_2Cl_8]^{3-}$ anion, a derivative of the Tc_2^{5+} core[1]. However, in 1980, Preetz and Peters[9] were successful in preparing $(Bu_4N)_2Tc_2Cl_8$, together with $(Bu_4N)_3Tc_2Cl_8$, by a procedure that involved the mossy zinc reduction of $(NH_4)_2TcCl_6$ in aqueous HCl followed by cation exchange using Bu_4NCl. The green complex $(Bu_4N)_2Tc_2Cl_8$ can be converted to the bromide derivative $(Bu_4N)_2Tc_2Br_8$ by dissolving it in aqueous acetone/HBr[9]. The successful completion of an X-ray crystal structure determination on $(Bu_4N)_2Tc_2Cl_8$ provided incontrovertible proof as to the structure of the $[Tc_2Cl_8]^{2-}$ anion[10]. This salt is isostructural with $(Bu_4N)_2Re_2Cl_8$ and, like the latter, possesses a quadruply bonded dimetal unit with an eclipsed rotational conformation. While there is disorder associated with the orientation of the $[Tc_2Cl_8]^{2-}$ ions, the structure is of high precision; the Tc–Tc distance is 2.147(4) Å, this being the weighted average of Tc–Tc distances of 2.151(1) Å and 2.133(3) Å for the major and minor orientations[10]. One result that remains somewhat controversial is the electrochemical redox characteristics of the $[Tc_2Cl_8]^{2-}/[Tc_2Cl_8]^{3-}$ couple. It had previously been demonstrated[11] that in a mixed hydrochloric acid-ethanol solvent (1 : 9 by volume) the $[Tc_2Cl_8]^{3-}$ ion (as its yttrium salt) is reversibly oxidized to $[Tc_2Cl_8]^{2-}$ at + 0.14 V vs. SCE. On the other hand, solutions of $(Bu_4N)_2Tc_2Cl_8$ in CH_2Cl_2-0.1 M Bu_4NClO_4 are characterized by $E_{1/2}$ = – 0.13 V vs. SCE at a rotating platinum electrode[10], so that the electrochemical potential for this process is solvent dependent; it is possible that solvolysis of the $[Tc_2Cl_8]^{3-}$ ion occurs in HCl-EtOH solutions (*vide infra*).

The Tc–Tc bond distance of 2.147 Å in $(Bu_4N)_2Tc_2Cl_8$ has itself posed an interesting dilemma, namely, why is this distance longer than the Tc-Tc distances reported for the $[Tc_2Cl_8]^{3-}$ ion in its ammonium and potassium salts (2.13(1) and 2.117(2)Å, respectively)[12,13]. This result is the opposite of that which might be expected based upon a simple bond length-bond order correlation. To provide further supporting experimental evidence, the crystal structure of the yttrium salt $Y[Tc_2Cl_8] \cdot 9\ H_2O$ has been determined[14]. Once again, the Tc–Tc distance (2.105(1) Å) was found to be shorter than in the $[Tc_2Cl_8]^{2-}$ ion. The explanation for this phenomenon is provided in Sect. 5.6.

While quadruply bonded carboxylate-bridged dirhenium(III) complexes of the type $Re_2(O_2CR)_4X_2$ are easily prepared and have been thoroughly characterized, comparable ditechnetium(III) carboxylate compounds are still very rare[1]. The pivalate derivative $Tc_2(O_2CCMe_3)_4Cl_2$ was prepared in low yield and structurally characterized a few years back[15]. More recently, the diamagnetic acetate complex $Tc_2(O_2CCH_3)_4Cl_2$ was prepared as cherry red crystals from the reaction between $KTcO_4$ and hydrochloric and acetic acids in a hydrogen atmosphere[16]. The product yield was not reported. A comparison between the IR spectra and X-ray diffraction powder patterns of $Tc_2(O_2CCH_3)_4Cl_2$ and $Re_2(O_2CCH_3)_4Cl_2$ has led to the suggestion that these complexes are isostructural.

The autoclave reaction (120 °C, 30 atm) between $K_3Tc_2Cl_8 \cdot 2\ H_2O$ and glacial acetic acid in an atmosphere of argon or hydrogen has been used to prepare the crystalline Tc_2^{5+} derivatives $Tc_2(O_2CCH_3)_4Cl$ (green) and $K[Tc_2(O_2CCH_3)_4Cl_2]$ (pale brown), admixed with K_2TcCl_6 and a material purported, but not proven, to be a Tc(II) complex[17]. Unfortunately, the product yields were not reported, and the crystals in the mixture were apparently separated by hand or by use of a solvent. Various other autoclave reductions have been carried out using mixtures containing $MTcO_4$, M_2TcX_6, $M_3Tc_2Cl_8 \cdot 3\ H_2O$, or M_2TcOCl_5 (M = NH_4 or K; X = Cl, Br, or I) and concentrated HX in an H_2 atmosphere, and these have been described as affording the brown or black crystalline Tc(II) compounds $M_2Tc_2X_6 \cdot 2\ H_2O$ (X = Cl or Br) and $(TcX_2 \cdot 0.5\ H_2O)$ (X = Br or I)[18]. The correct formulation of these materials remains to be proven. Whereas the chloride salts "$M_2Tc_2Cl_6 \cdot 2\ H_2O$" react with hot hydrochloric acid to give $[Tc_2Cl_8]^{3-}$, the bromides give solutions containing $[TcBr_6]^{2-}$ when they are dissolved in hydrobromic acid in the presence of air[18]. The reaction between "$M_2Tc_2Br_6 \cdot 2\ H_2O$" and glacial acetic acid in an argon atmosphere (50 atm) at 230–250 °C gives green crystalline $Tc_2(O_2CCH_3)_4Br$ (yield not specified)[18].

The complexes $Tc_2(O_2CCH_3)_4X$ (X = Cl or Br) and $K[Tc_2(O_2CCH_3)_4Cl_2]$ are clearly authentic derivatives of the Tc_2^{5+} core. They are paramagnetic and ESR-active, and possess magnetic moments in accord with the presence of a $\sigma^2\pi^4\delta^2\delta^{*1}$ ground state electronic configuration[17,19]. X-ray crystal structure determinations on $K[Tc_2(O_2CCH_3)_4Cl_2]$ and $Tc_2(O_2CCH_3)_4Cl$ have been completed[20,21]. The former salt contains the dinuclear $[Tc_2(O_2CCH_3)_4Cl_2]^-$ anion with Tc–Tc and Tc–Cl distances of 2.126 Å and 2.589 Å, respectively[20]. The complex $Tc_2(O_2CCH_3)_4Cl$ has a structure closely akin to that of $Tc_2(hp)_4Cl$ (hp is the anion of 2-hydroxypyridine)[22], in which chains of $[Tc_2(O_2CCH_3)_4]^+$ units are linked by bridging chloride ligands[21]. In all instances, the Tc–Tc distances of the acetate and hp derivatives of the Tc_2^{5+} core are shorter than the Tc–Tc distance in the quadruply bonded pivalate $Tc_2(O_2CCMe_3)_4Cl_2$ (2.192 Å); this trend mirrors that found for the $[Tc_2Cl_8]^{3-}$ and $[Tc_2Cl_8]^{2-}$ ions (*vide supra*). In passing, we should mention that a proposal has recently surfaced which attributes the decrease in Tc–Tc bond length on going from Tc_2^{6+} complexes to those of Tc_2^{5+} to an increase in Tc–Tc bond order from 4 to 4.5[23]. This interpretation is not supported by any meaningful theoretical calculations[1], and neither is there any experimental evidence which favors such an interpretation over the more commonly accepted one[10,14]. It must be emphasized that any simplistic interpretation which assumes that an *increase* in bond length will necessarily correlate with a *decrease* in bond order in multiply bonded dimetal systems is fraught with danger.

Spectrophotometric studies have been used to investigate the stability of solutions of the $[Tc_2Cl_8]^{3-}$ anion in hydrochloric acid[24,25]. At HCl concentrations below 3 M, hydrolysis to give $[Tc_2Cl_{8-n}(H_2O)_n]^{n-3}$ is said to occur, while for [HCl] > 3 M, disproportionation to generate $[TcCl_6]^{2-}$ (i.e. Tc(IV)) and Tc(II) species (identity unknown) is a dominant process. The presence of dissolved O_2 leads to oxidative cleavage of $[Tc_2Cl_8]^{3-}$ to give $[TcCl_6]^{2-}$ and $[TcO(OH)Cl_4]^{2-}$.

Dirhenium Compounds. In the case of dirhenium chemistry, the most striking advances have been encountered in the case of the so-called electron-rich Re–Re triple bonds ($\sigma^2\pi^4\delta^2\delta^{*2}$ electronic configuration) where a variety of new reactivity patterns have been established. Studies on such species constitute one of the most rapidly developing areas of multiple bond chemistry.

(a) Rhenium-Rhenium Quadruple Bonds ($\sigma^2\pi^4\delta^2$ Configuration). The quadruply bonded octahalodirhenate(III) anions and the halide-carboxylate complexes $Re_2(O_2CR)_4X_2$ and $Re_2(O_2CR)_2X_4$ continue to be the main focus of interest in the chemistry of the Re_2^{6+} core. Of special note has been the discovery of a new one-pot, high yield synthesis of $(Bu_4N)_2Re_2Cl_8$, the key starting material in multiple bond dirhenium(III) chemistry. The reaction of Bu_4NReO_4 with refluxing benzoyl chloride at ~210 °C has provided a means whereby $(Bu_4N)_2Re_2Cl_8$ can be prepared easily, quickly, and in high (90%) yield[26]. In view of the high cost of rhenium (as perrhenate) and the rather low overall yield in which $(Bu_4N)_2Re_2Cl_8$ is produced by other methods[1], this constitutes a very important development in dirhenium(III) chemistry. An alternative starting material for the synthesis of $(Bu_4N)_2Re_2Cl_8$ is $ReOCl_3(PPh_3)_2$[26], a complex which is itself prepared from M^IReO_4 in high yield. It is believed that this reaction proceeds via the intermediacy of $Re_2(O_2CPh)_2Cl_4$; thus the role of PhCOCl is to reduce and chlorinate the metal centers and then couple them via the agency of benzoate bridges[26].

Further refinements have been made in the assignment of the electronic absorption spectrum of the $[Re_2Cl_8]^{2-}$ ion, especially with regard to the identification of the weak, forbidden bands that lie between the $\delta \rightarrow \delta^*$ transition (~14,200 cm^{-1}) and the intense near-UV absorptions[27]. These assignments have been aided by an MO calculation using the relativistic SCF-Xα-SW method[27], while details of another theoretical study have been reported by Hay[28] that involve multiconfiguration valence bond calculations in conjunction with relativistic effective core potentials. Of further importance insofar as the electronic structures and spectroscopic properties of the $[Re_2X_8]^{2-}$ ions (X = F, Cl, Br, or I) are concerned, have been the results of a very thorough resonance Raman spectral study[29]. It was found that with excitation within the contours of the $\delta \rightarrow \delta^*$ and $X(\pi) \rightarrow \delta^*$ electronic transitions, progressions in ν_1 (i.e. ν(Re–Re)) and ν_2 (i.e. ν_{sym}(Re–Cl)), respectively, dominate the RR spectrum of each ion[29].

An interesting facet of the photochemistry of $[Re_2Cl_8]^{2-}$ involves the electron-transfer chemistry of the luminescent excited state $[Re_2Cl_8]^{2-*}$, which is an $\delta\delta^*$ singlet[30,31].

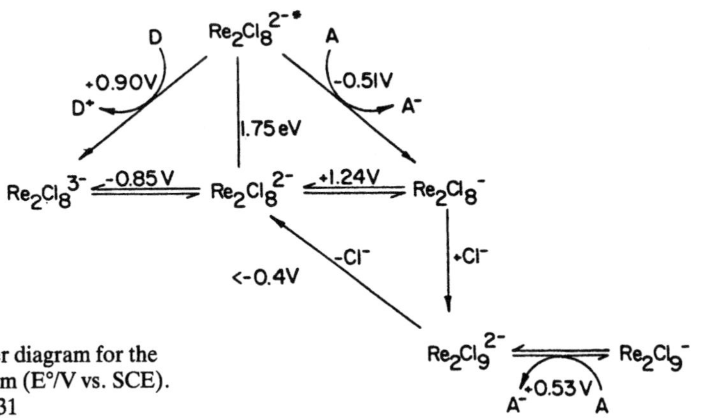

Fig. 2-1. Modified Latimer diagram for the $[Re_2Cl_8]^{n-}/[Re_2Cl_9]^{n-}$ system (E°/V vs. SCE). By permission from Ref. 31

Various electron acceptors (e.g. TCNE) quench the $[Re_2Cl_8]^{2-*}$ luminescence in non-aqueous solvents to produce $[Re_2Cl_8]^-$ and the reduced acceptor. The luminescence is also quenched by secondary and tertiary aromatic amines (e.g. N,N,N′,N′-tetramethyl-*p*-phenylenediamine) in acetonitrile solution[30)]. Thus the δδ* singlet provides a facile route to the powerful oxidant $[Re_2Cl_8]^-$, a species that has its own interesting chemistry. For example, it reacts with Cl^- to generate the well known mixed-valence species $[Re_2Cl_9]^{2-}$ [1)]; this can in turn be oxidized to $[Re_2Cl_9]^-$ by a suitable acceptor[31)]. This transformation corresponds to a net two-electron photochemical oxidation of $[Re_2Cl_8]^{2-}$ to $[Re_2Cl_9]^-$ [31)]. The potentials for the $[Re_2Cl_8]^{n-}$ and $[Re_2Cl_9]^{n-}$ systems that have been measured or estimated from the results of various spectroscopic and electrochemical experiments can be shown most appropriately in a modified Latimer diagram (Fig. 2-1)[31)].

Several X-ray crystal structures have been determined for salts that contain the $[Re_2Cl_8]^{2-}$ and $[Re_2Br_8]^{2-}$ anions, all of which were published pre-1982[1, 8)]. The only recent structure determination that has been carried out is that of the salt $[(DMA)_2H]_2Re_2Br_8$ (DMA = dimethylacetamide)[32)], but details of the structure are not yet available. Of the octahalodirhenate(III) anions, the crystal structures of the fluoride and iodide salts $(Bu_4N)_2Re_2X_8$ remain to be determined. However, there can be no doubt that both anions possess the typical $[Re_2X_8]^{2-}$ structure, with an eclipsed rotational conformation and a very short Re–Re bond. In this regard, the isolation of the novel compound $Re_4I_8(CO)_6$ is of special relevance[33)]. It is prepared by the I_2 oxidation of $Re_2I_2(CO)_6(THF)_2$ in heptane, and has a structure that can be viewed formally as involving interaction between two $[Re(CO)_3]^+$ fragments and $[Re_2I_8]^{2-}$; this association occurs via bent Re–I–Re bridges that involve six of the iodine atoms of $[Re_2I_8]^{2-}$. As a consequence of this interaction, which gives rise to $[(CO)_3ReI_3]$ units, the rotational conformation within the Re_2I_8 unit is staggered rather than eclipsed and, as a result, the Re–Re bond length is longer (2.279(1) Å)[33)] than that found in eclipsed $[Re_2X_8]^{2-}$ species (X = Cl or Br) where the distance is typically 2.22–2.25 Å[1)]. In accord with the usual bonding scheme for Re_2L_8 species[1)], as the conformation changes from eclipsed to staggered, the δ bonding diminishes (i.e. the d_{xy}-d_{xy} overlap) and the metal-metal bond order decreases from four to three.

The quadruply bonded dirhenium(III) carboxylates $Re_2(O_2CR)_4X_2$, $Re_2(O_2CR)_3X_3$, and $Re_2(O_2CR)_2X_4$ (X = Cl, Br or I), and other closely allied species, have a well developed and documented structural and reaction chemistry[1]. No major advances in this chemistry have been reported during the last four years, although there have been some refinements in our knowledge of these systems. The synthesis and thermal behavior of the hydrates $Re_2(O_2CCH_3)_2X_4 \cdot 2\,H_2O$ (X = Cl or Br) have been the subject of another report. Their preparation involves the high pressure hydrogen reduction of $KReO_4$ in a mixture of HX and acetic acid[34]. These hydrated complexes react with ligands (L) such as py, DMF, DMSO, dimethylacetamide, and Ph_3PO to form adducts of the type $Re_2(O_2CCH_3)_2X_4L_2$[35]. Previous X-ray crystal structure determinations[1,8] have

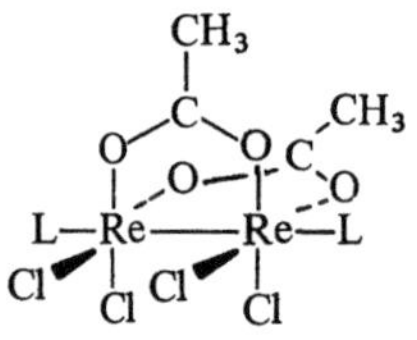

shown that in these adducts (including the hydrates) there is a cisoid arrangement of carboxylate groups with the ligands L being bound in the axial sites (colinear with the Re–Re bond). A close similarity in the spectral properties (IR and electronic absorption) of this group of acetate complexes[35] implies that they are closely related structurally. Several mixed halide derivatives of the type $Re_2(O_2CCH_3)_2Cl_{4-x}Br_xL_2$ (L = DMF, DMSO, or dimethylacetamide) were obtained serendipitously as by-products during the autoclave synthesis of *cis*-$Re_2(O_2CCH_3)_2Br_4L_2$[36]. It appears that the presence of chloride in the products arose from impurities that had been adsorbed on the wall of the autoclave from previous experiments that had involved an HCl-containing reaction mixture[36].

Thermal studies have shown[34,37] that the axial ligands L of $Re_2(O_2CCH_3)_2X_4L_2$ can be lost on heating, although it is clear that this can be a complex process. However, in the case of L = H_2O, the anhydrous acetates $Re_2(O_2CCH_3)_2X_4$ are formed following loss of L[34]. These same anhydrous compounds have also been formed through the thermal decomposition of $Re_2(O_2CCH_3)_4X_2$[34]. Note that the decomposition of the pivalate complex $Re_2(O_2CCMe_3)_4Cl_2$ in vacuo gives green $Re_2(O_2CCMe_3)_2Cl_4$ or pink $Re_2(O_2CCMe_3)_3Cl_3$ depending upon the decomposition temperature[38]. X-ray structural studies on anhydrous $Re_2(O_2CCH_3)_2X_4$ (X = Cl or Br) have shown that a *trans*-arrangement of carboxylate groups is present[39], thereby demonstrating that loss of the axially bound water molecules is accompanied by a *cis* → *trans* isomerization. The Re–Re bond distances are 2.211(3) Å for X = Cl and 2.216(3) Å for X = Br; these molecules possess essentially the same molecular structure as that of the previously characterized $Re_2(O_2CCMe_3)_2Cl_4$ and $Re_2(O_2CPh)_2I_4$[1]. Neighboring dinuclear units are linked via weak Re–X···Re bridges (at 3.09 Å). A useful reaction of $Re_2(O_2CCH_3)_2Br_4$ is its conversion to $Re_2(O_2CCH_3)_3Br_3$ upon treatment with a stoichiometric quantity of acetic acid in chloroform[34]. In an excess of glacial acetic acid, $Re_2(O_2CCH_3)_2Br_4$ is converted to $Re_2(O_2CCH_3)_4Br_2$. Under similar conditions, $Re_2(O_2CCH_3)_2Cl_4$ dissolves in an excess of cold glacial acetic acid to give violet $Re_2(O_2CCH_3)_3Cl_3$[34].

A variety of dirhenium(III) carboxylates of the types $Re_2(O_2CCH_3)_4Br_2$, *trans*-$Re_2(O_2CCH_3)_2X_4$ (X = Cl or Br), *cis*-$Re_2(O_2CCH_3)_2X_4L_2$ (L = py, DMSO, etc.), $Re_2(O_2CH)_2Cl_4L_2$ (L = DMF or DMSO), and $(NH_4)_2[Re_2(O_2CH)_2Cl_6]$ have been studied by ^{35}Cl or ^{81}Br NQR spectroscopy and the results compared with data for salts containing the $[Re_2Cl_8]^{2-}$ and $[Re_2Br_8]^{2-}$ anions[40)]. Also of note are some screening studies that have been carried out on the use of $[Re_2(O_2CC_2H_5)_4]SO_4$ and several derivatives of the type $Re_2(O_2CR)_2X_4 \cdot 2\ H_2O$ as anti-tumor agents[41)].

For the first time, a detailed electronic absorption spectroscopic study on a derivative of the type $Re_2(O_2CR)_4X_2$ has been reported. Single crystal polarized absorption spectra have been obtained for the pivalate complex $Re_2(O_2CCMe_3)_4Cl_2$ from 15,000 to 35,000 cm^{-1} [42)]. The electric-dipole-allowed $\delta \rightarrow \delta^*$ transition ($^1A_{1g} \rightarrow {}^1A_{2u}$) has been assigned to a peak at 20,600 cm^{-1} which exhibits molecular z polarization. This assignment accords with previous conclusions based upon solution measurements on various $Re_2(O_2CR)_4X_2$ complexes (X = Cl, Br, or I)[43)]. A much weaker band in the spectrum of $Re_2(O_2CCMe_3)_4Cl_2$ at 16,500 cm^{-1} with x,y polarization has been attributed to the spin forbidden $^1A_{1g} \rightarrow {}^3A_{2u}$ ($\delta \rightarrow \delta^*$) transition[42)].

(b) Rhenium-Rhenium Triple Bonds ($\sigma^2\pi^4\delta^2\delta^{*2}$ Configuration). The single most important group of complexes that possess this configuration are those of stoichiometry $Re_2X_4(PR_3)_4$ (X = Cl, Br, or I)[1)]. Several important advances in their redox and reaction chemistry have been made over the last four years. First, however, we note that a synthesis of $Re_2Br_4(PMe_3)_4$ has been reported[44)] that involves the reaction of PMe_3 with $Re_3Br_9(THF)_3$ in THF. This procedure is very similar to one that has been employed previously to obtain complexes of this type, viz., the reductive cleavage of Re_3Cl_9 or Re_3Br_9 by monodentate tertiary phosphine in acetone or alcohol solvents[45, 46)].

Relativistic Xα-SW calculations on the hypothetical molecule $Re_2Cl_4(PH_3)_4$ have demonstrated[47)] that the two electrons in excess of those necessary to ensure a quadruple bond reside in the Re–Re δ^* orbital. An analysis of the low temperature electronic absorption spectra (at 5 K) of $Re_2Cl_4(P\text{-}n\text{-}Pr_3)_4$ and its one-electron oxidized congener $[Re_2Cl_4(P\text{-}n\text{-}Pr_3)_4]PF_6$, possessing a $\sigma^2\pi^4\delta^2\delta^{*1}$ configuration, supports this bonding picture[47)]. In the case of the paramagnetic cation, the $\delta \rightarrow \delta^*$ transition is observed in the near-IR region of the spectrum (at ~6600 cm^{-1}). This band exhibits a vibrational progression of frequency 275 ± 5 cm^{-1}; this suggests that the Re–Re bond in this system is stronger than in $[Re_2Cl_8]^{2-}$, even though the bond order is less.

The redox behavior of $Re_2X_4(PR_3)_4$, together with that of compounds of the types $Re_2X_5(PR_3)_3$ and $Re_2X_6(PR_3)_2$, accords very nicely with the notion that these Re_2^{n+} species are representatives of the $\sigma^2\pi^4\delta^2$ (n = 6), $\sigma^2\pi^4\delta^2\delta^{*1}$ (n = 5), and $\sigma^2\pi^4\delta^2\delta^{*2}$ (n = 4) configurations[1)]. Recent work has further exploited these findings, particularly with regard to the development of reagents for bringing about one-electron oxidations or reductions of these compounds. Thus, $NOPF_6$ has been used to oxidize $Re_2Cl_4(PMe_2Ph)_4$ in two one-electron steps to give $[Re_2Cl_4(PMe_2Ph)_4]PF_6$ and $[Re_2Cl_4(PMe_2Ph)_4](PF_6)_2$[48)], while cobaltocene has proven to be a very effective reducing agent in the conversion of $Re_2Cl_6(PR_3)_2$ ($PR_3 = PEt_3$, P-*n*-Pr_3, $PMePh_2$, or $PEtPh_2$) to paramagnetic $[(\eta^5\text{-}C_5H_5)_2Co]$-$[Re_2Cl_6(PR_3)_2]$[49)]. The oxidation of $Re_2Cl_5(PMePh)_3$ to $[Re_2Cl_5(PMePh_2)_3]PF_6$ by $NOPF_6$ and its reduction to $[(\eta^5\text{-}C_5H_5)_2Co][Re_2Cl_5(PR_3)_3]$ have also been accomplished[49)]. Note that these redox reagents have been successfully used in related systems, such as in the cobaltocene reductions of the dirhenium(III) carboxylates $Re_2(O_2CR)_4Cl_2$, where R = C_3H_7, CMe_3, or Ph, and the 2-hydroxypyridine (Hhp) derivatives $Re_2(hp)_4X_2$, where X = Cl, Br, or I, to give the paramagnetic salts $[(\eta^5\text{-}C_5H_5)_2Co][Re_2(O_2CR)_4Cl_2]$ and $[(\eta^5\text{-}C_5H_5)_2Co][Re_2(hp)_4X_2]$, which are derivatives of the Re_2^{5+} core[49, 50)]. The carboxylate anions had previously only been generated electrochemically[43]. The preceding reactions, when combined with observations from previous investigations and incorporated into the redox scheme shown below, demonstrate the scope of the redox chemistry of the mixed halide-phosphine complexes of Re_2^{n+} (n = 6, 5, or 4). This scheme is an updated version of a similar one that was described in an earlier review[51].

Redox Scheme for Re_2^{n+} Complexes

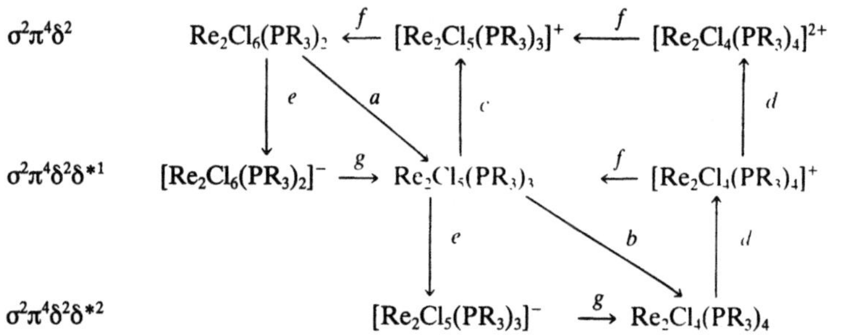

Chemical reactions are as follows: (a) reaction with PR_3 at room temperature; (b) reaction with PR_3 under reflux; (c) reaction with $NO^+PF_6^-$ in CH_2Cl_2 at room temperature; (d) reaction with $NO^+PF_6^-$ in CH_3CN at room temperature; (e) reaction with Cp_2Co in acetone at room temperature; (f) reaction with one-equivalent of Cl^-; (g) reaction with one equivalent of PR_3.

The isolation and structural characterization of the complexes $Re_2Cl_4(PMe_2Ph)_4$, $[Re_2Cl_4(PMe_2Ph)_4]PF_6$, and $[Re_2Cl_4(PMe_2Ph)_4](PF_6)_2$ have provided us with the first opportunity to probe the structural changes that take place in a series of complexes which possess M–M bond orders of 4, 3.5, and 3 and identical sets of monodentate ligands[48]. The consequences of this, insofar as changes in the Re–Re bond are concerned, are discussed in Sect. 5.6 of this review.

Let us now turn our attention to mixed halide-phosphine complexes of the Re_2^{4+} core that contain bidentate phosphine/arsine ligands. The compounds $Re_2X_4(LL)_2$, where X = Cl, Br, or I, and LL = 1,2-bis(diphenylphosphino)ethane (dppe) or 1-diphenylphosphino-2-diphenylarsinoethane (arphos), are especially interesting and well characterized examples[1,7]. These compounds, like their $Re_2X_4(PR_3)_4$ analogues, exhibit two accessible one-electron oxidations. They possess structures in which the ligands LL bridge the two

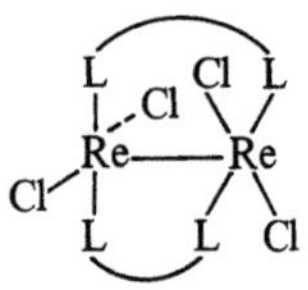

(L–L = dppe or arphos)

metal centers thereby conferring a staggered rotational geometry. Note that there are other multiply bonded dimetal complexes of this type (for Mo and W)[1], and they constitute a fascinating class of chiral compounds (see Sect. 5.7). When an additional methylene bridge is inserted between the P and/or As donor atoms of such bidentate ligands, i.e. as in LL = $Ph_2P(CH_2)_3PPh_2$, then the resulting complex $Re_2Cl_4[Ph_2P(CH_2)_3PPh_2]_2$ is one in which the two ligands LL each *chelate* to a single metal atom and the conformation is eclipsed[52].

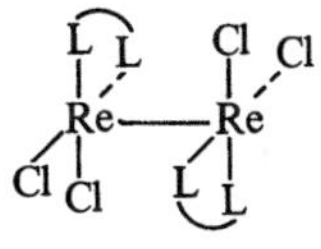

Although the ligand $Ph_2PCH_2PPh_2$(dppm) can chelate, it shows a propensity for bridging two metal atoms[53]. This is the situation in the case of $Re_2Cl_4(dppm)_2$, a complex that can be prepared most conveniently from the reaction between $Re_2Cl_6(P\text{-}n\text{-}Bu_3)_2$ and dppm in methanol[54]. The "mixed-valent" species $Re_2Cl_5(dppm)_2$ results when diethyl ether is used as the reaction solvent[54]. The structure of $Re_2Cl_4(dppm)_2$ has recently been solved and shown to resemble that of its dimolybdenum(II) analogue $Mo_2Cl_4(dppm)_2$[55], with the two bridging dppm ligands in a *trans*-disposition to one another; the Re–Re distance is 2.238 Å[56].

A study of the electrochemical properties of $Re_2Cl_4(dppm)_2$ has established that it exhibits two reversible one-electron oxidations in CH_2Cl_2 solutions with Bu_4NPF_6 as supporting electrolyte[54]. The reaction of Cl^- with the one-electron oxidation product $[Re_2Cl_4(dppm)_2]^+$ ($E_{1/2}$ = +0.27 V vs. SCE) generates $Re_2Cl_5(dppm)_2$. If the latter complex is in turn oxidized to $[Re_2Cl_5(dppm)_2]^+$, which can be accomplished at a potential of ca. +0.6 V vs. SCE, and this oxidation product reacted with Cl^-, then the dirhenium(III) complex $Re_2Cl_6(dppm)_2$ is formed. The latter compound possesses a μ-dichloro-bridged structure (Fig. 2-2) with a Re–Re double bond[54]. The latter conclusion is based upon the observed Re–Re bond length (2.616(1) Å)[54], together with the results of a theoretical treatment[57] which concluded that such a species should possess a $\sigma^2\pi^2\delta^{*2}\delta^2$ electronic configuration. The conversion of $Re_2Cl_4(dppm)_2$ to $Re_2Cl_5(dppm)_2$ and $Re_2Cl_6(dppm)_2$ can, therefore, be represented as follows:

$$\mathrm{Re_2Cl_4(dppm)_2} \xrightarrow[+\mathrm{Cl^-}]{-e^-} \mathrm{Re_2Cl_5(dppm)_2} \xrightarrow[+\mathrm{Cl^-}]{-e^-} \mathrm{Re_2Cl_6(dppm)_2}$$

	$\sigma^2\pi^4\delta^2\delta^{*2}$	$\sigma^2\pi^4\delta^2\delta^{*1}$	$\sigma^2\pi^2\delta^{*2}\delta^2$
Bond order	3	3.5	2

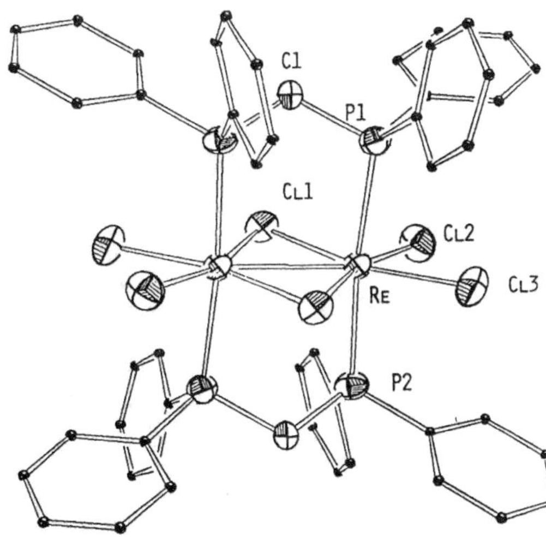

Fig. 2-2. The structure of $Re_2Cl_6(dppm)_2$. By permission from Ref. 54

The conversion of $Re_2Cl_5(dppm)_2$ to $Re_2Cl_6(dppm)_2$ represents a novel example of a reaction in which the oxidation of the Re_2^{5+} core to Re_2^{6+} results in a *reduction* in the Re–Re bond order rather than an increase to a bond of order 4. Incidentally, $Re_2Cl_6(dppm)_2$ can be prepared directly from the $[Re_2Cl_8]^{2-}$ ion through the reaction of the latter with dppm in dichloromethane[54]. When a primary alcohol ROH is used as solvent, this reaction affords an alkoxide complex of stoichiometry $Re_2Cl_5(OR)(dppm)_2$ (R = CH_3, C_2H_5, or C_3H_7) which possesses a structure closely akin to that of $Re_2Cl_6(dppm)_2$, with one of the terminal Re–Cl bonds of the latter complex replaced by an Re–OR unit[54].

With the use of the bridging bidentate ligand bis(diphenylphosphino)amine (dppa) in place of dppm, the complexes $Re_2Cl_6(dppa)_2$, $Re_2Cl_4(dppa)_2$, and $Re_2Cl_4(PMePh_2)_2(dppa)$ have been prepared[54]. Their spectroscopic and electrochemical properties show them to be structurally similar to their dppm analogues[1, 54]. With acetonitrile as the reaction solvent, the ligand 2-(diphenylphosphino)pyridine (Ph_2Ppy) reacts with $[Re_2Cl_8]^{2-}$ to give the complex $Re_2(\mu\text{-}Cl)_2(\mu\text{-}Ph_2Ppy)_2Cl_4$, which is believed to have a structure like that of $Re_2Cl_6(dppm)_2$ and $Re_2Cl_6(dppa)_2$[58].

Actually, the reactions between the Ph_2Ppy ligand and dirhenium complexes have proven to be unusually diverse and have led to the discovery of some new facets of multiple bond chemistry. In addition to the formation of $Re_2Cl_6(Ph_2Ppy)_2$ as outlined above, the reactions of Ph_2Ppy with the dirhenium(III) complexes $(Bu_4N)_2Re_2Cl_8$ and $Re_2Cl_6(PR_3)_2$, where R = Et or *n*-Bu, can result in the reduction of the dimetal core[58, 59]. Thus, the compounds $Re_2Cl_6(PR_3)_2$ are converted into $Re_2Cl_4(Ph_2Ppy)_2(PR_3)$, where R = Et or *n*-Bu, when acetone is the reaction solvent. However, the most unusual reaction is that between $(Bu_4N)_2Re_2Cl_8$ and Ph_2Ppy in methanol which first produces $Re_2Cl_4(Ph_2Ppy)_3$, a species which then eliminates HCl to give the *ortho*-metallated complex $Re_2Cl_3(Ph_2Ppy)_2[(C_6H_5)(C_6H_4)Ppy]$[58, 59]. This was the first example of an *ortho*-metallation reaction occurring at a multiple bond in a molecule of the M_2L_8 type. Soon thereafter a second example was encountered involving the *ortho*-metallation of PPh_3 by the diosmium(III) compounds $Os_2(O_2CR)_4Cl_2$ (see Sect. 4). The structure of $Re_2Cl_3(Ph_2Ppy)_2[(C_6H_5)(C_6H_4)Ppy]$ is shown in Fig. 2-3.

Another product from the same reaction that affords $Re_2Cl_3(Ph_2Ppy)_2$-$[(C_6H_5)(C_6H_4)Ppy]$ is the dirhenium(II) salt $[Re_2Cl_2(Ph_2Ppy)_4]Cl_2$. It can be metathesized

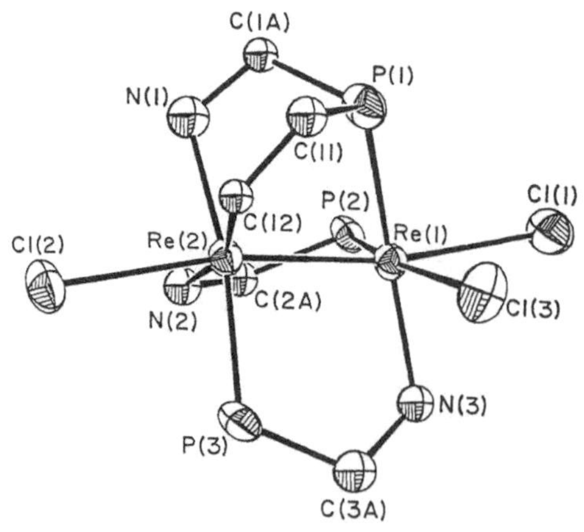

Fig. 2-3. The structure of the central portion of the compound $Re_2Cl_3(Ph_2Ppy)_2$-$[(C_6H_5)(C_6H_4)Ppy]$ with the phenyl rings omitted. By permission from Ref. 58

with KPF_6 to give $[Re_2Cl_2(Ph_2Ppy)_4](PF_6)_2$, a compound whose structure has been determined by X-ray crystallography[58]. This complex represents only the second example of a multiply bonded dimetal unit bridged by four *neutral* bridging ligands. The only other example is the isoelectronic dimolybdenum(I) complex $Mo_2Cl_2[F_2PN(CH_3)PF_2]_4$[60].

The well behaved redox chemistry of $Re_2Cl_4(dppm)_2$ in Bu_4NPF_6–CH_2Cl_2 can be contrasted with the behavior observed when acetonitrile (and other nitriles) is used as the solvent. This complex reacts rapidly and irreversibly with RCN ligands to form complexes of stoichiometry $[Re_2Cl_3(dppm)_2(NCR)_2]X$ (X = Cl^- or PF_6^-; R = CH_3, C_2H_5, C_6H_5, or 4-Ph–C_6H_4)[61]. The use of $^{31}P\{^1H\}$ NMR spectroscopy to monitor the formation of $[Re_2Cl_3(dppm)_2(NCC_2H_5)_2]Cl$ in the reaction between $Re_2Cl_4(dppm)_2$ and propionitrile shows that these reactions occur in a two step fashion in which a RCN ligand first coordinates to one of the Re atoms of $Re_2Cl_4(dppm)_2$ to generate a 1 : 1 adduct, possibly possessing the molecular A-frame type structure $(C_2H_5CN)ClRe(\mu\text{-}Cl)(\mu\text{-}dppm)_2ReCl_2$, followed by the addition of a second nitrile ligand to the same Re atom with concomitant loss of Cl^- to generate $[(C_2H_5CN)_2ClRe(\mu\text{-}dppm)_2ReCl_2]^+$[61].

(c) Reactions of Rhenium-Rhenium Triple and Quadruple Bonds with π-Acceptor Ligands. The cleavage of multiply bonded dimetal species by the π-acceptor ligands CO, NO, and the alkyl and aryl isocyanides, is a reaction course that is commonly encountered and which can be used to advantage in the synthesis of many mononuclear complexes[1,51]. Recent examples include the carbonylation of $[Re_2Cl_4(PMe_2Ph)_4]^{n+}$, where n = 0, 1 or 2, to give the carbonyl derivatives $ReCl(CO)_3(PMe_2Ph)_2$, $ReCl(CO)_2(PMe_2Ph)_3$, and/or $ReCl_3(CO)(PMe_2Ph)_3$[62], and the reaction of $(Bu_4N)_2Re_2Cl_8$ in acetonitrile with CO at 100 atm. and ~90 °C to give $ReCl(CO)_5$ as the major product and the ionic compound $[cis\text{-}Re(NCCH_3)_4(CO)_2]_2ReCl_6$ as a minor one[63]. In the case of the alkyl isocyanide ligands RNC, the very stable homoleptic species $[Re(CNR)_6]^+$ (R = CMe_3 or C_6H_{11}) can be obtained in good yield from $Re_2(O_2CR)_4Cl_2$ (R = Me or Ph)[64]. When the triply bonded dirhenium(II) complexes $Re_2Cl_4(PR_3)_4$ are treated with these same isocyanides, then a similar reaction course ensues to give cationic complexes of stoichiometry $[Re(CNR)_4(PR_3)_2]^+$ which can be isolated as their PF_6^- salts[64]. In contrast to these results, the more "halide-rich" complexes $(Bu_4N)_2Re_2X_8$ (X = Cl or Br) and $Re_2Cl_6(PEtPh_2)_2$ give the mononuclear rhenium(III) complexes $[Re(CNR)_5X_2]PF_6$ (X = Cl or Br) and $[Re(CNR)_4(PEtPh_2)Cl_2]PF_6$, respectively, in which reduction to Re(I) has not occurred[64]. Similar behavior has been encountered in the reactions of aryl isocyanides ArNC (Ar = phenyl, *p*-tolyl, 2,6-dimethylphenyl, and 2,4,6-trimethylphenyl) with the quadruply bonded complexes $Re_2(O_2CCH_3)_4Cl_2$ and $(Bu_4N)_2Re_2Cl_8$. In refluxing methanol, these reactions provide an excellent high yield synthetic route to $[Re(CNAr)_6]PF_6$[65]. When the reactions between ArNC and $(Bu_4N)_2Re_2Cl_8$ are conducted at room temperature, then Re(III)-containing intermediates of the types $[Re(CNAr)_6]_2Re_2Cl_8$, $[Re(CNAr)_6][ReCl_4(CNAr)_2]$, and $ReCl_3(CNAr)_3$ can be isolated[65].

While the cleavage reactions described in these most recent reports[62–65], together with the results from earlier studies[51], have been of considerable synthetic value, relatively little mechanistic information has yet been forthcoming because these reactions are quite facile and are difficult to control. However, the dppm-bridged complex $Re_2Cl_4(dppm)_2$ has a much greater stability towards cleavage by the π-acceptor ligands CO, RNC, and ArNC, and very recent findings have established[66,67] that multiply-bonded dinuclear complexes which contain these ligands can be isolated. We are optimistic that a study of these systems will provide us with insights into the mechanisms of these reactions. Included among the species that have been isolated and structurally characterized to date are the novel unsymmetrical dicarbonyl complex $Cl_2Re(\mu\text{-}Cl)(\mu\text{-}CO)(\mu\text{-dppm})_2ReCl(CO)$[66], and the A-frame molecule $(RNC)ClRe(\mu\text{-}Cl)(\mu\text{-dppm})_2ReCl_2$[67]. These results appear to offer some exciting prospects for the development of further novel dirhenium chemistry.

3 Detailed Update of Ditungsten(II) Chemistry

The chemistry of quadruply bonded compounds of the Group VI elements Cr, Mo, and W has been notable for the difficulty that has been experienced in preparing certain key ditungsten(II) species, specifically, the tetracarboxylates $W_2(O_2CR)_4$ and salts of the octachloroditungstate(II) anion $[W_2Cl_8]^{4-}$. This contrasts with the ease of preparing the analogous dimolybdenum(II) compounds[1]. The latter are quite stable and they display an extensive reaction chemistry that has been of paramount importance in the development of the chemistry of the Mo_2^{4+} core. While advances have continued to be made in our understanding of dimolybdenum chemistry, we will concentrate here on the less numerous, although equally spectacular, developments in ditungsten chemistry.

Ditungsten(II) Carboxylates. Following several earlier unsuccessful attempts to prepare an authentic tetracarboxylate of ditungsten(II), a report describing the isolation and structural characterization of the trifluoracetate $W_2(O_2CCF_3)_4$ was finally published in mid-1981[68]. This synthesis was achieved by the Na/Hg reduction of the ditungsten(III) complex $W_2Cl_6(THF)_4$ in tetrahydrofuran at $-20\,°C$, followed by the addition of CF_3CO_2Na (reaction (3-1)). A blue THF-soluble intermediate (vide infra) which is formed in this reaction is unstable and decomposes (slowly at 0 °C, more rapidly at room temperature) in the absence of CF_3CO_2Na. While the yield of $W_2(O_2CCF_3)_4$ obtained

$$W_2Cl_6(THF)_4 + 2\ Na/Hg + 4\ NaO_2CCF_3 \xrightarrow{THF} W_2(O_2CCF_3)_4 \qquad (3\text{-}1)$$

by this method is quoted as being quite good, an alternative (and preferred) synthetic procedure involves the Na/Hg reduction of a mixture of WCl_4 and sodium trifluoroacetate in cold (0 °C) THF[69]. Other alkyl carboxylate derivatives can apparently be prepared by this procedure[69], and this method also constitutes a convenient route to the corresponding aryl carboxylate derivatives $W_2(O_2CAr)_4$ (Ar = Ph, C_6H_4-*p*-OMe, and C_6H_2-2,4,6-Me_3)[70, 71]. A method that also seems to be of general use for the synthesis of alkyl carboxylate derivatives involves the reaction at room temperature of hydrocarbon solutions of 1,2-$W_2Et_2(NMe_2)_4$ with acid anhydrides $(RCO)_2O$, where R = Me, Et, or CMe_3, as shown in reaction (3-2)[72]. Product yields after recrystallization are in the range

$$W_2Et_2(NMe_2)_4 + 4\,(RCO)_2O \rightarrow W_2(O_2CR)_4 + 4\,RCONMe_2 + C_2H_4 + C_2H_6 \qquad (3\text{-}2)$$

40–65%.

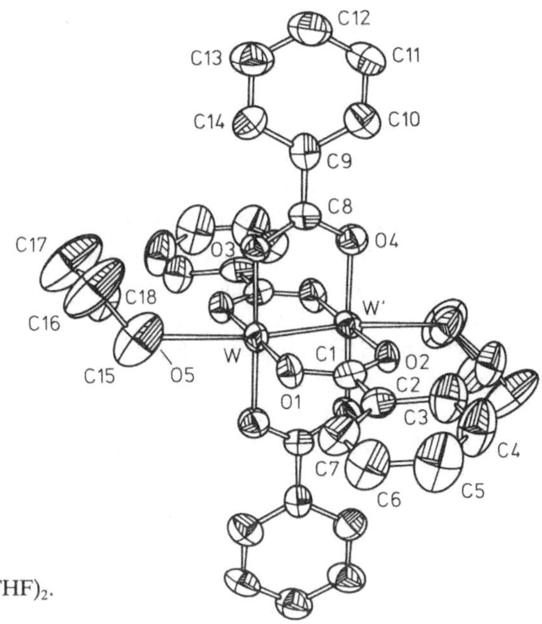

Fig. 3-1. The structure of $W_2(O_2CPh)_4(THF)_2$. By permission from Ref. 71

The air-sensitive trifluoroacetate complex is, like the other alkyl carboxylate derivatives, sublimable, and an intense parent ion multiplet is seen in its mass spectrum, thereby providing conclusive evidence that the dinuclear complex remains intact in the vapor phase. Another important property of the alkyl and aryl carboxylate complexes is the ease with which they form adducts with various Lewis bases; in this regard their properties resemble those of the corresponding dimolybdenum(II) carboxylates[1]. Several of these adducts can be obtained in suitable crystalline form, and single crystal X-ray structure analyses have confirmed their identity as axial 1:2 adducts (see, for example, Fig. 3-1). Only in the case of the bis-toluene solvate of $W_2[O_2CC_6H_2\text{-}2,4,6\text{-}Me_3]_4$, in which the toluene molecules are not bound to the metal atoms, has the structure proven to be that of a ditungsten(II) carboxylate in which significant axial interactions are *absent*. In the case of $W_2(O_2CEt)_4$, intermolecular W–O axial interactions (2.665(4) Å) link the dinuclear units into infinite chains. A summary of the W–W distances for the structurally characterized derivatives is given in Table 3-1; note that the shorter W–W distance in $W_2(O_2CC_6H_2\text{-}2,4,6\text{-}Me_3)_4$ can be ascribed to the absence of axial ligands, a

Table 3-1. W–W Distances in ditungsten(II) carboxylates

Compound	r(W–W), Å	Ref.
$W_2(O_2CEt)_4$	2.189(1)	72
$W_2(O_2CCF_3)_4(diglyme)_{2/3}$	2.209(2)	69
$W_2(O_2CCF_3)_4(PPh_3)_2$	2.242(2)	69
$W_2(O_2CPh)_4(THF)_2$	2.196(1)	71
$W_2(O_2CC_6H_4\text{-p-OMe})_4(THF)_2 \cdot 2$ THF	2.203(1)	71
$W_2[O_2CC_6H_2\text{-}2,4,6\text{-}Me_3]_4 \cdot 2$ toluene	2.176(1)	71

structural effect that is seen in other tetracarboxylate dimetal(II) complexes of the Group VI elements[1)]. This "weakening" of the W–W bond by axial W–L interactions is also reflected by changes in the Raman-active ν(W–W) mode. We here refer to a shift in ν(W–W) from 310 cm^{-1} in $W_2(O_2CCF_3)_4$ to 280 cm^{-1} in $W_2(O_2CCF_3)_4(PPh_3)_2$[69)]. This is very similar to that found between $Mo_2(O_2CCF_3)_4$ (r(Mo–Mo) = 2.090(4) Å, ν(Mo–Mo) = 397 cm^{-1}) and $Mo_2(O_2CCF_3)_4(py)_2$ (r(Mo–Mo) = 2.129(2) Å, ν(Mo–Mo) = 363 cm^{-1})[1)]. Other noteworthy spectroscopic studies that have been carried out on these ditungsten(II) carboxylates include the ^{19}F and ^{183}W NMR of $W_2(O_2CCF_3)_4$[69)], and the assignment of the $\delta \rightarrow \delta^*$ transition in the case of the aryl carboxylate derivatives[71)]. The latter band occurs in the region between 450 and 600 nm, and the observed peak energies correlate quite nicely with those attributed to the $\delta \rightarrow \delta^*$ transition of the analogous dimolybdenum(II) complexes. Finally, the UV PE spectrum of $W_2(O_2CCF_3)_4$ (and its molybdenum analogue) has been measured[73, 74)], but the assignments that have been made have generated some controversy. The δ ionizations of $Mo_2(O_2CCF_3)_4$ and $W_2(O_2CCF_3)_4$ appear at 8.76 and 7.39 eV, respectively, a difference which correlates with the much greater susceptibility of ditungsten(II) complexes to oxidation and oxidative-addition reactions. While this assignment seems to be indisputable, the suggestion[73, 74)] that the π and σ ionizations occur very close together (9.01 and ~9.76 eV) has been challenged[75)]. It seems very unlikely that such a close coincidence will occur; the σ ionization can be expected to be assignable to a higher energy feature in the spectrum.

In contrast to the reaction of PPh_3 with $W_2(O_2CCF_3)_4$, which leads to a 1:2 adduct containing axially bound phosphine ligands[69)], the behavior with other phosphine ligands (PMe_3, PEt_3, and P-*n*-Bu_3) is more complicated than this[76)]. Toluene solutions of $W_2(O_2CCF_3)_4$ react with these three trialkylphosphines to yield red to red-orange air-sensitive 1:2 adducts. The ^{19}F and $^{31}P\{^1H\}$ NMR spectra (+25 to −50 °C) support the presence of a single isomer in which the phosphine ligands are bound equatorially and two bridging bidentate and two monodentate trifluoroacetate ligands are present[76)]. A single crystal X-ray structure determination on $W_2(O_2CCF_3)_4(P\text{-}n\text{-}Bu_3)_2$ has confirmed that this is the case in the solid state as well (Fig. 3-2); the W–W bond length in this complex is 2.224(1) Å[76)]. Thus, like its dimolybdenum analogue[77–79)], $W_2(O_2CCF_3)_4$ can react with phosphine ligands to give axial (class I) or equatorial (class II) adducts.

The reaction of $W_2(O_2CCF_3)_4$ or $W_2(O_2CCF_3)_4(PMe_3)_2$ with an excess of PMe_3 gives the dark green tris(phosphine) adduct $W_2(O_2CCF_3)_4(PMe_3)_3$, which is stable in solution at room temperature but dissociates a molecule of PMe_3 to give $W_2(O_2CCF_3)_4(PMe_3)_2$ when heated in benzene[76)]. The ^{19}F and $^{31}P\{^1H\}$ NMR spectra of this complex show that it has the same structure in solution as that determined in the solid state by X-ray crystallography (Fig. 3-3). A comparison of the spectroscopic properties of $W_2(O_2CCF_3)_4(PMe_3)_3$ with those of the previously reported molybdenum complex $Mo_2(O_2CCF_3)_4(PMe_3)_3$[77)], suggests that these complexes are isostructural. However, the differences that do exist in solution between $Mo_2(O_2CCF_3)_4(PR_3)_2$ and $W_2(O_2CCF_3)_4(PR_3)_2$, and between $Mo_2(O_2CCF_3)_4(PMe_3)_3$ and $W_2(O_2CCF_3)_4(PMe_3)_3$, have been linked to differences in the M–P bond strengths (W–P > Mo–P)[76)].

Unlike the dimolybdenum(II) tetracarboxylates, (especially $Mo_2(O_2CCH_3)_4$), which have served as the key starting materials in the development of multiply bonded dimolybdenum chemistry[1)], the related ditungsten(II) species $W_2(O_2CR)_4$ have not yet found a similar use. This reflects their ease of oxidation, with consequent difficulty of handling. This is especially true for those reactions in which oxidative addition can occur across the

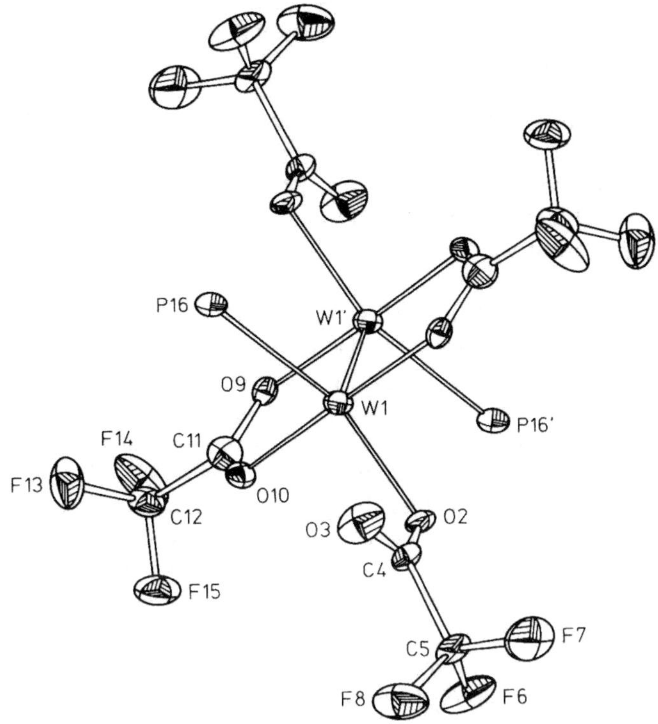

Fig. 3-2. The structure of $W_2(O_2CCF_3)_4(P\text{-}n\text{-}Bu_3)_2$. By permission from Ref. 76

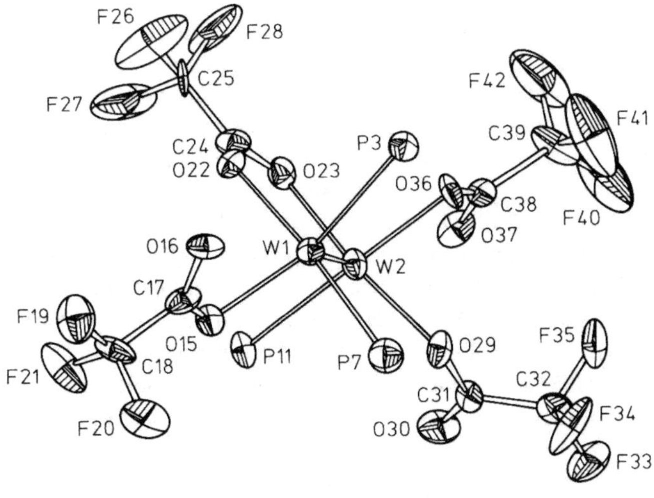

Fig. 3-3. The structure of $W_2(O_2CCF_3)_4(PMe_3)_3$. By permission from Ref. 76

W–W multiple bond. Indeed, it is generally true that complexes which contain the W_2^{4+} core are unstable in this regard, a point that will receive further consideration a little later in this section.

The Octachloroditungstate Anion and Derivatives Thereof. The second key ditungsten(II) species whose preparation has over the years proven to be something of a challenge is the octachloroditungstate(II) anion, $[W_2Cl_8]^{4-}$. While an SCF-Xα-SW calculation had in 1977 indicated that this anion should be capable of existence[80)], several earlier attempts to prepare it were successful. However, a variety of phosphine-containing derivatives were prepared and structurally characterized in the process[1)]. These compounds were of the types $W_2Cl_4(PR_3)_4$ (PR_3 represents a monodentate tertiary phosphine ligand), α-$W_2Cl_4(dmpe)_2$, and the α- and β-forms of $W_2Cl_4(dppe)_2$ (dmpe = $Me_2PCH_2CH_2PMe_2$; dppe = $Ph_2PCH_2CH_2PPh_2$), all of which have well characterized analogues in dimolybdenum(II) chemistry. The preparative details and spectroscopic properties of these complexes are available[81, 82)], together with the results of X-ray structural studies on $W_2Cl_4(PMe_3)_4$, α-$W_2Cl_4(dmpe)_2$, and α- and β-$W_2Cl_4(dppe)_2$[83–85)]; the α- and β-isomers differ in terms of whether they contain chelating (α-) or bridging (β-) $R_2PCH_2CH_2PR_2$ ligands. In these phosphine-containing complexes the W–W bond lengths are in the range 2.26 to 2.31 Å, and are thus somewhat longer than those found for the ditungsten(II) tetracarboxylates (Table 3-1) and other complexes that contain four monoanionic bridging ligands[1)]. Note that the complex that possesses the longest W–W distance is β-$W_2Cl_4(dppe)_2$, a compound that by virtue of the bridging dppe ligands that span the ditungsten unit is forced to adopt a partially staggered rotational conformation (it is twisted 31.3° from eclipsed); this leads to a weakening of the δ-bond and thus accounts, in part, for the lengthening of the W–W bond.

One reaction of note for complexes of the type $W_2Cl_4(PR_3)_4$ is their one-electron oxidation to generate $[W_2Cl_4(PR_3)_4]^+$. This process approaches electrochemical reversibility, as judged by cyclic voltammetric measurements, the $E_{1/2}$ value for a solution of $W_2Cl_4(P\text{–}n\text{-}Bu_3)_4$ in Bu_4NPF_6–CH_2Cl_2 being − 0.24 V vs SCE[82)]. This behavior parallels that found for the dimolybdenum analogues $Mo_2Cl_4(PR_3)_4$[86)], although the ditungsten compounds are, as expected, much more easily oxidized. The complex $W_2Cl_4(P\text{–}n\text{-}Bu_3)_4$ has been oxidized chemically to the paramagnetic, ESR-active salt $[W_2Cl_4(P\text{–}n\text{-}Bu_3)_4]PF_6$ using $[Ag(NCMe)_4]PF_6$ as the oxidant[82)]. While very few studies have yet been carried out that utilize $W_2Cl_4(PR_3)_4$ as starting material for the synthesis of other W_2^{4+} complexes, future work along these lines can be expected. The conversion of W_2Cl_4-$(P\text{–}n\text{-}Bu_3)_4$ to $W_2Cl_4(bpy)_2$ (bpy = 2,2′-bipyridine) provides one such recent example[87)].

The development of successful synthetic routes to the compounds $W_2Cl_4(PR_3)_4$ has led to a method whereby salts of the $[W_2Cl_8]^{4-}$ anion can be isolated. One of the procedures that has been used to prepare $W_2Cl_4(PR_3)_4$ involves the Na/Hg reduction of WCl_4 in THF in the presence of PR_3[81, 82)]. In the *absence* of PR_3, the reaction proceeds first to give green $W_2Cl_6(THF)_4$, which is in turn reduced further by Na/Hg to generate an intense blue colored species. Work-up of this solution at 0 °C has afforded $Na_4(THF)_xW_2Cl_8$ as a reactive blue powder[82)]. Since there is insufficient chloride ion in this synthesis to convert all the $W_2Cl_6(THF)_4$ to $Na_4(THF)_xW_2Cl_8$, the reaction is believed to proceed as in (3-3).

$$W_2Cl_6(THF)_4 + 2\ Na/Hg \xrightarrow[-30°]{THF} \tfrac{1}{2} Na_4(THF)_xW_2Cl_8 + \text{“}WCl_2\text{”} \qquad (3\text{-}3)$$

In accord with this, the yield of $Na_4(THF)_xW_2Cl_8$ never exceeds 50%; "WCl_2" represents the proportion of the tungsten that is chloride-deficient. Attempts to increase the yield of $[W_2Cl_8]^{4-}$ by carrying out the Na/Hg reduction of WCl_4 in the presence of added chloride were unsuccessful[82]. In one such instance, $[(Ph_3P)_2N]Cl$ was used as the chloride source and this produced $[(Ph_3P)_2N]_2W_2Cl_9$ as dark violet crystals[88].

The THF of solvation in $Na_4(THF)_xW_2Cl_8$ can be replaced by bidentate ethers and amines, e.g. dimethoxyethane (DME) and tetramethylethylenediamine (TMEDA), to give salts such as $Na_4(DME)_4W_2Cl_8$ and $Na_4(TMEDA)_4W_2Cl_8$, all of which display a band at ~600 nm in their electronic absorption spectra that is assignable to the $\delta \rightarrow \delta^*$ transition of the $[W_2Cl_8]^{4-}$ anion[82,89]. A crystal structure determination on $Na_4(TMEDA)_4W_2Cl_8$ has confirmed the existence of the $[W_2Cl_8]^{4-}$ anion in this salt[89]. The W–W distances of 2.259(1) and 2.254(1) Å that were found for the two independent $[W_2Cl_8]^{4-}$ ions in the unit cell[89] are essentially the same as the W–W distance determined for $W_2Cl_4(PMe_3)_4$[83].

Samples of $Na_4(THF)_xW_2Cl_8$ react with phosphines (PMe_3 and $P\text{-}n\text{-}Bu_3$) to give $W_2Cl_4(PR_3)_4$ in essentially quantitative yield[82]. This salt also reacts[82] with 6-methyl-2-hydroxypyridine (Hmhp) in the presence of Et_3N to give the known quadruply bonded ditungsten(II) complex $W_2(mhp)_4$[1]. A similar reaction with a THF solution that contained pivalic acid and triethylamine at −30 °C has been used to obtain the pivalate complex $W_2(O_2CCMe_3)_4$ as a yellow powder. However, attempts to extend this procedure to give a pure sample of the acetate complex $W_2(O_2CCH_3)_4$ were unsuccessful[82].

One of the most interesting reactions discovered in dimolybdenum(II) chemistry is the conversion of quadruply bonded $Mo_2X_4L_4$ (X = Cl or Br; L = CH_3OH, C_2H_5CN, THF, or PR_3) to the tetranuclear clusters $Mo_4X_8L_4$. These compounds possess a rectangular geometry and can be regarded as tetrametal analogues of cyclobutadiyne[90,91]. Schematically this addition of two M≡M units one to the other can be represented as follows.

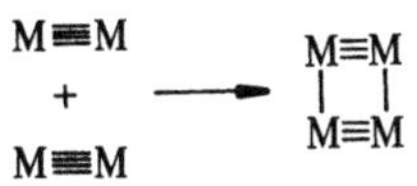

This chemistry has now been extended to tungsten through the sequence of reactions (3-4)–(3-6) which utilize the synthetic strategies that have been developed by Schrock and co-workers[81,82], in the preparation of $W_2Cl_4(PR_3)_4$.

$$2\ WCl_4 + 2\ Na/Hg \xrightarrow{THF} W_2Cl_6(THF)_4 + 2\,NaCl \qquad (3\text{-}4)$$

$$W_2Cl_6(THF)_4 + 2\ P\text{-}n\text{-}Bu_3 \xrightarrow{THF} W_2Cl_6(THF)_2(P\text{-}n\text{-}Bu_3)_2 + 2\,THF \qquad (3\text{-}5)$$

$$2\ W_2Cl_6(THF)_2(P\text{-}n\text{-}Bu_3)_2 + 4\ Na/Hg \xrightarrow{THF} W_4Cl_8(P\text{-}n\text{-}Bu_3)_4 + 4\,NaCl + 4\,THF \qquad (3\text{-}6)$$

An X-ray crystal structure determination on this complex has shown[90] that in its essential features it is similar to the structure of $Mo_4Cl_8(PEt_3)_4$, with the short and long W–W distances being 2.309(2) and 2.840(1) Å, respectively. By using the approach represented in reaction (3-7), the mixed metal cluster $Mo_2W_2Cl_8(PMe_3)_4$ has been prepared[92].

$$2\ MoW(O_2CCMe_3)_4 + 4\ PMe_3 + 8\ AlCl_3 \xrightarrow[\text{reflux}]{\text{THF}} Mo_2W_2Cl_8(PMe_3)_4 + 8\ AlCl_2(O_2CCMe_3) \qquad (3\text{-}7)$$

This product presumably consists of one of the following isomers,

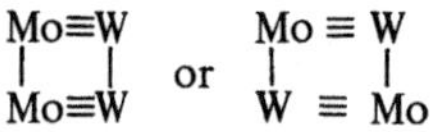

or a mixture of them; at the present time it is not known which is the case[92].

An important comparative study has been made of the Mo–Mo and W–W quadruple bonds in $M_2Cl_4(PMe_3)_4$ using SCF-Xα-SW calculations and PE spectroscopy[93]. Section 5.5 should be consulted for further details.

Other Ditungsten(II) Complexes. The 2-oxopyridine type ligand I, and related monoanionic ligand bridges such as 2-amino-6-methylpyridine and 2,4-dimethyl-6-hydroxypyrimidine have played an important role in the development of the chemistry of quad-

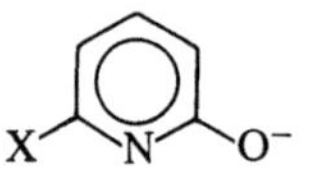

Ia (X = H), hp^- Ic (X = Cl), chp^-
Ib (X = CH_3), mhp^- Id (X = F), fhp^-

ruply bonded dimetal complexes, including those of ditungsten[1]. While the complexes $W_2(mhp)_4$ and $W_2(chp)_4$ were synthesized and structurally characterized prior to 1981, the related derivative $W_2(fhp)_4$ (Id) has been isolated much more recently. It is prepared as its 1 : 1 adduct with THF by the Na/Hg reduction of a THF solution containing Na(fhp) and WCl_4[94]. This complex, together with its Cr and Mo analogues, has an unprecedented

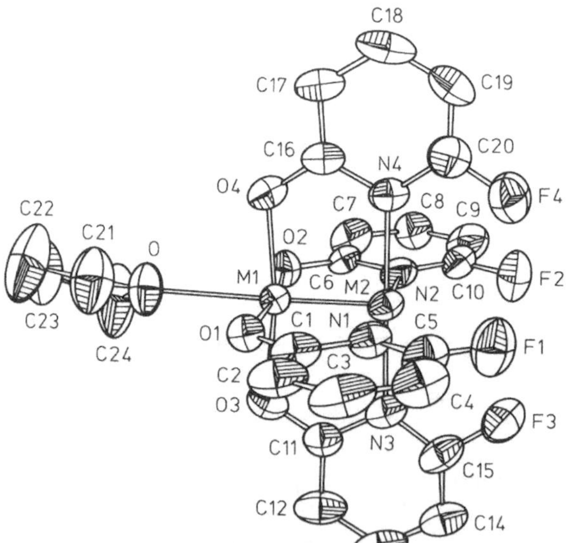

Fig. 3-4. The structure of $W_2(fhp)_4(THF)$. By permission from Ref. 94

molecular structure in which the four bridging fhp ligands are all orientated in the same direction (Fig. 3-4); a THF molecule is coordinated axially to the metal atom that is coordinated entirely by oxygen atoms[94]. A synthetic procedure similar to that used to prepare $W_2(fhp)_4$ has been adapted for the synthesis of $W_2(PhNpy)_4$ (PhNpy is 2-anilinopyridine), a complex that contains an eclipsed $[W_2N_8]$ core[95]. Spectroscopic studies on $W_2(fhp)_4$ and $W_2(PhNpy)_4$, and on related ditungsten(II) complexes that contain four monoanionic bridging ligands, have provided good evidence that the $\delta \rightarrow \delta^*$ transition is located between 500 and 600 nm[94, 95].

Reactivity of the Tungsten-Tungsten Quadruple Bond. Spectroscopic (PES and electronic absorption) and structural evidence supports the notion that the δ-component of the W–W quadruple bond is significantly weaker than that in the case of M≡Mo. This is also reflected in the much greater ease of oxidizing $W_2Cl_4(PR_3)_4$ to $[W_2Cl_4(PR_3)_4]^+$ than $Mo_2Cl_4(PR_3)_4$ to $[Mo_2Cl_4(PR_3)_4]^+$ [82, 86]. An important consequence of this is the ease with which quadruply bonded ditungsten(II) complexes can be oxidized to ditungsten(III), or even ditungsten(IV), species. This was first recognized in the case of the oxidation of $W_2(mhp)_4$ by gaseous HX (X = Cl or Br) to give $[W_2Cl_9]^{3-}$ and $W_2X_4(OR)_4(ROH)_2$ (R = Me or Et)[96, 97]. A very similar situation holds for the $[W_2Cl_8]^{4-}$ ion when its solutions are treated with HCl. The $[W_2Cl_9]^{3-}$ ion is formed quantitatively even at −78 °C, presumably via an intermediate such as $[W_2Cl_9H]^{4-}$ or $[W_2Cl_8H]^{3-}$ [82]. A further interesting example is the reaction of $W_2Cl_4(P\text{-}n\text{-}Bu_3)_4$ with two equivalents of benzoic acid which gives rise to the ditungsten(III) complex $W_2(\mu\text{-}H)(\mu\text{-}Cl)Cl_2(O_2CPh)_2(P\text{-}n\text{-}Bu_3)_2$. This complex possesses a structure that is comprised of two edge sharing octahedra[98]. Note that when $W_2Cl_4(P\text{-}n\text{-}Bu_3)_4$ is reacted with an excess of acetic acid a much more complicated oxidation occurs to give the triangular tungsten(IV) complex $W_3(\mu_3\text{-}O)(\mu_2\text{-}O)_3(O_2CCH_3)Cl_5(P\text{-}n\text{-}Bu_3)_3$[82, 99].

As is the case for many other multiply bonded dimetal species[1], the W≡W bond is subject to cleavage by π-acceptor ligands. Recent examples of this include the reductive cleavage of $W_2(mhp)_4$ and $W_2(dmhp)_4$ (mhp and dmhp are the anions of 2-hydroxy-6-methylpyridine and 2,4-dimethyl-6-hydroxypyrimidine, respectively) by phenyl isocyanide to give $W(CNPh)_6$ in fair yield[100], and the conversion of $W_2Cl_4(PR_3)_4$ (R = Me or *n*-Bu) to $W(CO)_3Cl_2(PR_3)_2$ and *trans*-$W(CO)_4(PR_3)_2$ by carbon monoxide at atmospheric pressure in toluene at 70 °C[101]. The former reaction[100] provides a good synthetic route to the otherwise difficult to synthesize $W(CNPh)_6$.

4 Detailed Update of Diruthenium and Diosmium Chemistry

Since completion of the pertinent sections of "Multiple Bonds Between Metal Atoms" there have been notable discoveries in the chemistry of Ru_2^{n+} and Os_2^{n+} compounds. Present knowledge of this chemistry is several times as great as before. It may be noted that there is still no evidence for Fe–Fe bonding of a comparable type, i.e., involving oxidation numbers of +2 or greater. For example, a recent report on compounds of the type $[FeX(ArCO_2)]_x$ indicates that they all have magnetic moments, per Fe, of 4.8–5.3 BM[102]. This shows that each metal atom has a high-spin d^6 electron configuration and rules out the presence of metal-metal bonded diiron units.

Diruthenium(II,II) Compounds. In 1981 Collman and coworkers[103] made the remarkable discovery that pyrolysis of mononuclear ruthenium(II) porphyrin dipyridine complexes, $Ru(P)py_2$ (P = octaethylporphyrin (OEP), *meso*-tetraphenylphorphyrin (TPP) or *meso*-tetra-p-tolylporphyrin (TTP)) gives rise to dinuclear (P)Ru=Ru(P) molecules, which were characterized by NMR and other physical methods, including a crystal structure of the $Ru_2(OEP)_2$ compound[104]. It has a Ru–Ru distance of 2.408(1) Å and can be assigned a metal-metal double bond based on the electron configuration $\sigma^2\pi^4\delta^2\delta^{*2}\pi^{*2}$. The porphyrin groups are slightly domed and each ruthenium atom lies about 0.30 Å out of the plane defined by the four coordinated nitrogen atoms in the direction of the other ruthenium atom. It was also found that by pyrolysis of an intimate mixture of two different $Ru(P)py_2$ compounds, mixed products, such as (OEP)RuRu(TTP) could be obtained.

These new porphyrin compounds of Ru_2^{4+} are comparable to the previously known[105] $Ru_2(C_{22}H_{22}N_4)_2$, in which the bond length was found to be 2.379(1) Å. There is only one other structurally characterized diruthenium(II,II) compound, namely, $Ru_2(mhp)_4$, where bridging by the 6-methyl-2-oxopyridine ligands occurs and the Ru–Ru distance is much shorter, 2.238(1) Å[106].

Diruthenium(II,III) Compounds. It has been one of the surprising and not convincingly explained features of diruthenium chemistry from the very beginning that the II,III mixed oxidation species tend to be the most stable and easily prepared compounds. In this early work, which is reviewed in detail in MBBMA, the compounds were all carboxylato-bridged ones. In all cases where measurements were made, the magnetic moments showed the presence of three unpaired electrons and theoretical studies showed that this could be attributed to a near degeneracy of the π^* and δ^* orbitals so that the electron configurations are $\sigma^2\pi^4\delta^2(\pi^*,\delta^*)^3$. Ru–Ru distances in all these carboxylates were in the range 2.248(1) Å to 2.290(1) Å. Electrochemical studies showed that these complexes possessed a quasireversible one-electron reduction, but no II,II product was ever isolated. More recently the preparation and structure of $Ru_2(O_2CH)_4Br$ have been reported[107].

Bear and Kadish and coworkers[108, 109] have shown that the $Ru_2(HNOCR)_4Cl$ compounds, with R = CF_3[108] and CH_3[109] can be prepared, and they have extensively studied the very complex electrochemical and spectroelectrochemical behavior of these compounds. These amidato compounds were obtained by heating $Ru_2(O_2CCH_3)_4Cl$ with an excess of the molten amide, whereby CH_3CO_2H is driven off and the products are obtained in excellent yields. No x-ray data have been reported for these compounds. They have magnetic moments at room temperature of 2.8–3.0 BM, which may indicate a doublet-quartet spin state equilibrium. The electrochemical behavior is very strongly dependent on the solvent used and on the presence of excess Cl^- ion. It is assumed that this is because of the coordination of solvent molecules or Cl^- ions in the axial positions of the $[Ru_2(HNOCR)_4]^+$ unit, whereby three types of complex, S–Ru–Ru–S$^+$, S–Ru–Ru–Cl and Cl–Ru–Ru–Cl$^-$ are possible.

For $Ru_2(HNOCCF_3)_4Cl$ in DMSO a single, essentially reversible, one-electron reduction occurs at -0.19 V; in less strongly coordinating solvents much more complicated behavior is observed. For $Ru_2(HNOCCH_3)_4Cl$ in DMSO two reductions are observed. The one at -1.22 V is suspected of being ligand-centered, while the one at -0.96 V is presumably analogous to the one observed for $Ru_2(HNOCCF_3)_4Cl$ at -0.19 V, namely a reduction of the Ru_2(II,III) core to a Ru_2(II,II) core. The large shift in potential is in keeping with the change from CF_3 groups to CH_3 groups. In addition, with $Ru_2(HNOCCH_3)_4Cl$ an oxidation wave, presumably corresponding to formation of the Ru_2(III,III) species $[Ru_2(HNOCCH_3)_4]^{2+}$ is seen at + 0.47 V. The process is fully reversible.

It has been shown that $Ru_2(O_2CCH_3)_4Cl$ also reacts with molten benzamide to afford $Ru_2(HNOCPh)_4Cl$, a brown microcrystalline solid, rather insoluble in common solvents[110]. In DMSO solution it exhibits a reduction presumed to be to the Ru_2(II,II) species at − 0.66 V. The most interesting thing about this compound, however, is the way in which it reacts with triphenylphosphine, to give a product with the following composition and structure:

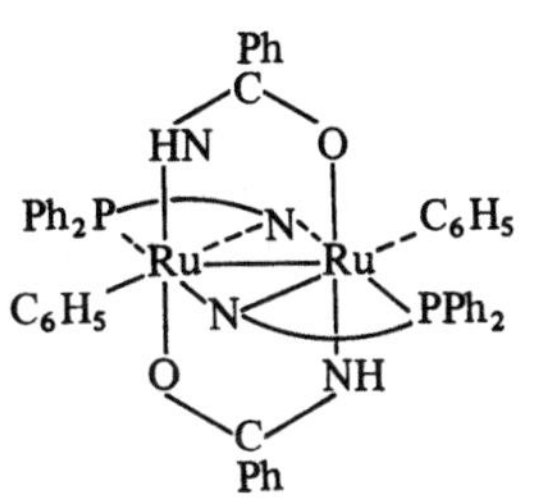

The bridging ligand represented by Ph_2P–N for the sake of clarity in the drawing is Ph_2P–O–C(Ph)–N. The overall structure is that of an edge-sharing bioctahedron with the N atoms of this ligand at the bridging positions. The Ru–Ru distance is 2.566(1) Å. This unusual ligand is (at least formally) the product of attack by PhC(NH)O on PPh_3, with the remarkable result that a phenyl group is transferred from the phosphorus atom to the ruthenium atom.

Two more reactions of $Ru_2(O_2CCH_3)_4Cl$ with molten reagents have more recently been carried out[111]. The reactions with 2-anilinopyridine and hydroxypyridine both lead

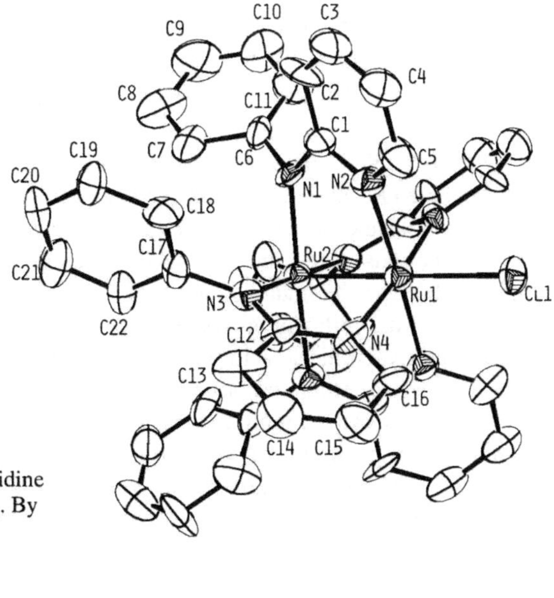

Fig. 4-1. The structure of the anilinopyridine diruthenium compound, $Ru_2Cl(PhNpy)_4$. By permission from Ref. 111

to products in which the bridging acetate ions are replaced by new bridging ligands and in both cases, these new ligands, which are unsymmetrical, all point in the same direction. The structure of the anilinopyridine product is shown in Fig. 4-1. As a result of having all four phenyl groups at one end of the Ru–Ru bond (which has a length of 2.275(3) Å) there is a twisting around the Ru–Ru bond and the Cl^- ion is necessarily coordinated at the other end. It is, in fact, very strongly coordinated, with a Ru–Cl distance of only 2.437(7) Å whereas all other such axial Ru–Cl bond lengths have been about 2.55 Å, or longer. In the hydroxypyridine product, $Ru_2Cl(hp)_4(Hhp)$, there is a neutral Hhp molecule, in its keto form, coordinated at one end (with an N–H···O hydrogen bond as well) and the Cl^- ion at the other. The Ru–Ru bond length here is 2.286(1) Å. Both compounds have magnetic moments indicative of three unpaired electrons.

Diruthenium(III,III) Compounds. Remarkable diruthenium(III,III) compounds, $Ru_2(CH_2XMe_3)_6$, (X = C, Si) have recently been reported by Wilkinson and coworkers[112]. They have obtained them by reaction of $Ru_2(O_2CCH_3)_4Cl$ with the neopentyl or $(CH_3)_3SiCH_2$ Grignard reagents and it appears that $Ru_2(O_2CCH_3)_2R_4$ molecules are intermediates. It is remarkable that oxidation of ruthenium occurs under these conditions that would normally be considered reducing. The $Ru_2(CH_2CMe_3)_6$ molecule has been characterized by X-ray crystallography. It has a staggered, ethane-like structure (similar to the triply-bonded molybdenum and tungsten M_2R_6 compounds), with an Ru–Ru distance of 2.311(3) Å. This work would appear likely to mark the opening of an important new phase of diruthenium chemistry.

Diosmium(III,III) Compounds. A few years ago in MBBMA it was correctly stated that of the two species, Ru_2^{6+} and Os_2^{6+}, "osmium alone has been unambiguously shown to form a triply bonded dimer, and even then there is only one such example." The example referred to is $Os_2(hp)_4Cl_2$, which had been structurally characterized in two solvated crystalline forms and shown to have a twisted rotational conformation with Os–Os distances of 2.344(2) Å and 2.357(1) Å[113]. There are now many Os_2^{6+} complexes containing a variety of ligands, as indicated in Table 4-1, where those that have been structurally characterized by X-ray crystallography are listed. The number of such structurally characterized Os_2^{6+} compounds is now literally an order of magnitude greater than when MBBMA was written.

(a) $Os_2(O_2CR)_4X_2$ Compounds. These were first reported in late 1981[114] and early 1982[115]. Three crystal structures have been established[115, 116] (see Table 4-1) and Fig. 4-2 shows the molecule of $Os_2(O_2CC_2H_5)_4Cl_2$, which is representative. In addition to the three compounds listed in Table 4-1, analogs with other carboxylic acids (e.g., $CClH_2CO_2H$) can be made. The first method of preparation was by reaction of $Na_2[OsCl_6]$ with a refluxing mixture of acetic acid and acetic anhydride to which a little concentrated hydrochloric acid has been added. After this procedure has been used to obtain the acetate, other compounds are then easily obtained by refluxing the acetate in other acids. It was later found that the use of $Na_2[OsCl_6]$ gave erratic results and another preparation of the acetate employing $[OsCl_6]^{2-}$ prepared from OsO_4 and HCl was recommended[117]. The acetate also reacts with excess 2-hydroxypyridine (Hhp) to give $Os_2(hp)_4Cl_2$; this compound was first prepared by reaction of $OsCl_3$ with Hhp.

Table 4-1. Compounds containing the diosmium(III,III) unit, Os_2^{6+}

Compound	Os–Os dist., Å	Os–Cl$_{ax}$ dist., Å	Color	Magnetic Moments, B.M.[a]		Ref.
$Os_2(hp)_4Cl_2 \cdot 2\ Et_2O$	2.344(2)	2.47, 2.50	Purple	2.04(300)	1.73(195)	113, 117
$Os_2(hp)_4Cl_2 \cdot 2\ CH_3CN$	2.357(1)	2.505(5)	Red			113, 117
$Os_2(O_2CCH_3)_4Cl_2$	2.314(1)	2.448(2)	Brown	1.65(288)	1.34(209)	116, 117
$Os_2(O_2CC_2H_5)_4Cl_2$	2.316(2)	2.430(5)	Brown	1.60(300)	1.44(188)	116, 117
$Os_2(O_2CC_3H_7)_4Cl_2$	2.301(1)	2.417(3)	Dk. green	1.63(300)	1.44(188)	117
$Os_2(O_2CCH_3)_2(Ph_2PC_6H_4)_2Cl_2$	2.271(1)	2.372(2)	Black	0.41(295)		118
$Os_2(O_2CC_2H_5)_2(Ph_2PC_6H_4)_2Cl_2$	2.272(1)	2.396(2)	Black	0.10(295)		118
$Os_2(PhCONH)_4Cl_2$	2.367(3)	2.47–2.51	Dk. green	1.76(295)		120
$Os_2(PhCONH)_4Br_2$	2.383(2)	2.59–2.63	Dk. green			120
$Os_2Cl_3(PhNpy)_3$	2.392(1)	2.449(5)	Dk. blue			121

[a] Temperature, K, in parentheses

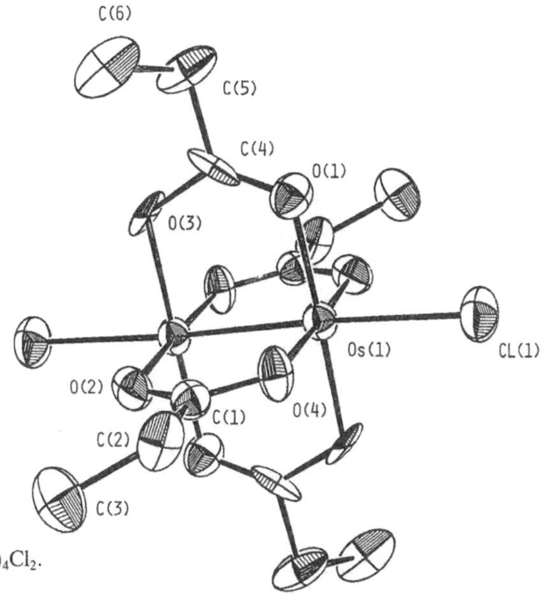

Fig. 4-2. The structure of $Os_2(O_2CC_2H_5)_4Cl_2$. By permission from Ref. 116

(b) Orthometallation of PPh_3 by $Os_2(O_2CR)_4Cl_2$ Compounds. One of the most remarkable reactions, Eq. 4-1, so far found for $Os_2(O_2CCH_3)_4Cl_2$ is that with triphenylphosphine in refluxing acetic acid, whereby two acetate groups are replaced by two orthometallated $Ph_2PC_6H_4$ ligands. When a similar reaction is carried out in propionic acid, the remaining acetate groups are replaced by propionate ions and the product is $Os_2(O_2CC_2H_5)_2(Ph_2PC_6H_4)_2Cl_2$[118]. The structure of the acetato molecule is shown in Fig. 4-3. Unfortunately, these compounds are extremely insoluble, which precludes a number of interesting further studies. However, the success of the reactions by which they are made may well be wholly, or partly, attributable to this insolubility, since the transfer of an orthohydrogen atom from a phenyl group to an acetate ion must in itself be

$$Os_2(O_2CCH_3)_4Cl_2 + 2\,Ph_3P \rightarrow Os_2(O_2CCH_3)_2(Ph_2PC_6H_4)_2Cl_2 + 2\,CH_3COOH \qquad (4\text{-}1)$$

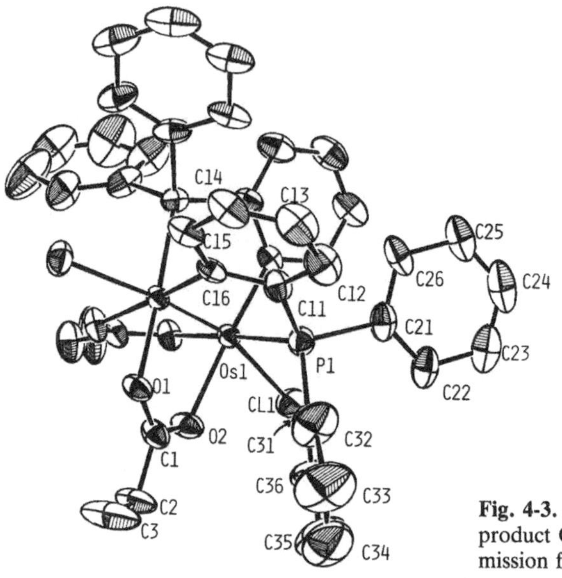

Fig. 4-3. The structure of the orthometallation product $Os_2(O_2CCH_3)_2(Ph_2PC_6H_4)_2Cl_2$. By permission from Ref. 118

quite unfavorable. The insolubility of the product could help to drive the reaction although another favorable factor is presumably the greater combined strength of the Os–P and Os–C bonds compared to the two Os–O bonds they replace. It is worth noting that an analogous reaction occurs with $Rh_2(O_2CCH_3)_4$ to give $Rh_2(O_2CCH_3)_2(Ph_2PC_6H_4)_2$[118, 119].

(c) Amidato Compounds, $Os_2(PhCONH)_4X_2$. By the reaction of $Os_2(O_2CCH_3)_4Cl_2$ with molten benzamide, $PhCONH_2$, an essentially quantitative yield of $Os_2(PhCONH)_4Cl_2$ is obtained[120], and this can be converted to $Os_2(PhCONH)_4Br_2$. The latter is the first compound of Os_2^{6+} that has a ligand other than Cl^- in the axial positions. Both compounds are air-stable, partially paramagnetic and rather poorly soluble except in DMSO.

(d) An Anilinopyridine Complex, $Os_2Cl_3(PhNpy)_3$. The reaction of $Os_2(O_2CCH_3)_4Cl_2$ with 2-(N-anilino)pyridine in presence of Me_3SiCl afforded an unusual unsymmetrical product, whose structure[121] is shown in Fig. 4-4. This is the only example of an Os_2^{6+} complex with a vacant axial position. It also has the longest known Os≡Os bond.

Diosmium(II,III) Compounds. The $Os_2(O_2CR)_4Cl_2$ and $Os_2(hp)_4Cl_2$ compounds have been shown electrochemically to undergo reversible one-electron reductions at potentials (vs SCE) of 0.30 ± 0.01 V for the carboxylates and 0.10 V for the hp compound[122]. These reductions can be accomplished chemically by using cobaltocene and solid products, $[C_{10}H_{10}Co][Os_2(O_2CR)_4Cl_2]$ and $[C_{10}H_{10}Co][Os_2(hp)_4Cl_2]$ can thereby be isolated. For the $[Os_2(O_2CR)_4Cl_2]^-$ ions effective magnetic moments (μ_{eff}, B.M.) are in the range 2.5–2.9, and they increase with decreasing temperature, while for $[Os_2(hp)_4Cl_2]^-$ an approximately constant value of μ_{eff} = 2.65 ± 0.03 B.M. was found between 308 K and 173 K. These moments, as well as some ESR data, are indicative of a quartet ground state with some thermal population of a higher doublet state. There are no structural data as yet for these II,III compounds.

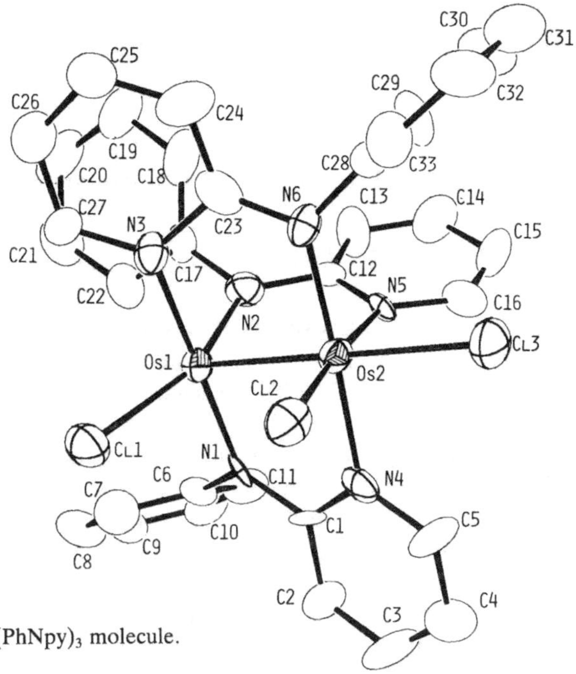

Fig. 4-4. The structure of the $Os_2Cl_3(PhNpy)_3$ molecule. By permission from Ref. 121

Diosmium(II,II) Compounds. The only known examples of these are porphyrin compounds, $Os_2(P)_2$, that are quite comparable to their ruthenium analogs[123]. They are prepared in a similar way, i.e., by pyrolysis of $Os(P)py_2$ compounds. Again, mixed ligand dimers can be obtained, such as (TTP)Os=Os(OEP), and in addition the mixed metal compound, (OEP)Ru=Os(OEP) has also been formed and identified by NMR, although not isolated. No structural data have yet been reported for an $Os_2(P)_2$ compound.

Electronic Structure and Bonding in the Diosmium Compounds. Although it might have naively been expected that the d^5–d^5 Os_2^{6+} complexes would behave electronically like their isoelectronic Re_2^{4+} analogs, this is not the case. The latter, e.g., compounds such as $Re_2Cl_4(PR_3)_4$, are diamagnetic and may be formulated as having $\sigma^2\pi^4\delta^2\delta^{*2}$ configurations with net triple bonds. Many of the osmium compounds have longer metal-to-metal bonds and have temperature-dependent paramagnetism, as first observed for the $Os_2(O_2CR)_4Cl_2$ compounds. Magnetic data are collected in Table 4-1. It is clear from a comparison of the magnetic data collected there and the Os–Os distances also listed in Table 4-1 that the paramagnetism increases as the Os–Os distance increases. It can also be seen that the magnetic moments decrease with decreasing temperature in each case where the temperature dependence has been measured.

The facts just summarized concerning the magnetic properties of the diosmium compounds are believed to be best explained by assuming that the d orbital overlap between the two metal atoms is such that a triplet state based on a $\sigma^2\pi^4\delta^2\delta^*\pi^*$ or $\sigma^2\pi^4\delta^2\pi^{*2}$ configuration lies close enough to the singlet ground state, based on the $\sigma^2\pi^4\delta^2\delta^{*2}$ configuration, to be thermally accessible at the temperatures of measurement shown in Table 4-1. At the shorter Os–Os distances, the singlet-triplet separation becomes greater and the paramagnetism is less. The orthometallated compounds, which have the shortest known Os–Os distances, are just barely paramagnetic.

5 Survey of Highlights in Other Areas

5.1 The M_2 Molecules

These molecules can be observed only in the gas phase or in inert gas matrices at very low temperatures. There has been little study of their reaction chemistry, which is, of course, difficult to do. In spite of these negative characteristics, the transition metal diatomic molecules, M_2, are of interest to chemists working on compounds containing M_2^{n+} cores because theoretical understanding of these "simple" M_2 species can help to further our understanding of the compounds containing M_2^{n+} cores. The word "simple" is appropriately placed in quotation marks because these molecules are "simple" only in chemical composition; their electronic structures have provided difficult challenges for the theoretician and experimental study requires sophisticated techniques. In any event, they afford the simplest structure circumstances under which strong, highly multiple bonds between metal atoms can be studied.

Considerable progress has been made in this area since the pertinent section of MBBMA was written. At that time only the Mo_2 molecule had been well characterized both experimentally and theoretically, although some results were available on other molecules that have only weak bonds in which the d orbitals play essentially no role. In such molecules, of which Fe_2, Co_2, Ni_2 and Cu_2 seem typical, the bond is formed by overlap of the valence shell $4s$ orbitals and the two d^n configurations then interact only weakly. However, these $d^n + d^n$ systems give rise to an enormous number of molecular electronic states close in energy to the ground state. These molecules are therefore very difficult to characterize definitively by either theoretical or experimental means.

However, our attention here will be directed to those molecules in which *d–d* bonding plays a significant part. We shall specifically discuss V_2, Cr_2, Mo_2, and Ru_2, though not in that order. Before taking up individual molecules, some generalizations are appropriate. The experimental study of M_2 molecules has been enormously benefited by the development of the supersonic free jet technique[1, 2)], whereby molecules in their ground states can be studied spectroscopically in the gas phase. In this way the bond length, force constant and electronic state can be determined with precision and certainty. In addition, of course, more traditional studies in inert gas matrices and by standard thermodynamic methods continue to be useful. From the theoretical side, computer codes for carrying out extensive configuration interaction (CI) calculations have become more efficient and local spin density (LSD) methods that go beyond the well-known Xα method for estimating exchange energies have been developed and applied.

The Mo_2 Molecule. This molecule has now been studied spectroscopically by the supersonic jet technique[126)]. The new results are more detailed and more precise than, but not essentially different from, those previously available and cited in MBBMA. The ground state is confirmed to be $^1\Sigma_g^+$ (meaning that all electrons are paired in fully occupied molecular orbitals). An electronic transition from this ground state to a $^1\Sigma_u^+$ state was found to occur without any appreciable weakening of the bond. It was suggested[126] that this can be understood by assuming that at the actual bond distance of ca. 1.94 Å, the bond is essentially a pentuple one, formed by the σ, 2π and 2δ components provided by d–d overlaps, while the sixth pair of electrons, occupying a $\sigma_g(5s)$ orbital, are effectively non-bonding. Excitation of one of these electrons (unless to an orbital with strongly bonding or antibonding character) therefore has a negligible effect on the properties of the molecule. These electrons were likened to electrons in a Rydberg type orbital, i.e., a large diffuse orbital lying outside the core of the molecule.

This proposal is not inconsistent with the theoretical work on the electronic structure of Mo_2 by Goodgame and Goddard[127]. These authors used a generalized valence bond (GVB) method of calculation in which electron correlation is, presumably, allowed for quite thoroughly. It is interesting that this calculation, which gives a potential energy curve having essentially correct values for r_0 and ν_0, shows a second very shallow minimum at about 3.1 Å, which the authors attribute to a state of the molecule in which there is *5s–5s* bonding but little or no *d–d* bonding.

The Cr_2 Molecule. Here we have the true *piece de la resistance* in this field, especially for theoreticians. Experimental work has now convincingly established[128–131] what Efremov and coworkers[132] had asserted nearly a decade earlier (on evidence that was not indisputable), namely that the Cr_2 molecule has a $^1\Sigma_g^+$ ground state with a very short, ca. 1.68 Å, distance. The vibrational frequency is 452 cm^{-1}.

This result has proved troublesome to account for theoretically. The same type of highly correlated GVB-MO calculation that Goodgame and Goddard used very successfully for Mo_2 led them to suggest[127], prior to the definitive measurements cited above, that Cr_2 would have an internuclear distance of 3.06 Å with very weak bonding (ca. 0.35 eV bond energy). There have since been several efforts to show why the Goodgame and Goddard calculation failed and to get the right answer. Kok and Hall[133] carried out a somewhat different type of highly correlated Hartree-Fock calculation and found a potential minimum at 1.73 Å, in fair agreement with experiment. They recognized the possibility that they may have had some good fortune in having had a cancellation of basis set errors and correlation errors, but they suggested that their method of calculation is inherently more likely to be successful with a molecule as compact and correlated as Cr_2. Another group[134] has done a Hartree-Fock-CI type calculation, with still other ways of choosing wave functions and configurations, but they failed to find any minimum at all, only a slight inflection. They nevertheless ventured to "conclude that the ground state... of Cr_2 (is) characterized by multiple 3d bonding".

Other workers[135] have used a so-called local spin density (LSD) method, in which certain assumptions are made about how to calculate the exchange and correlation energies. They obtained results for both Mo_2 and Cr_2 that agree very well with the experimental data (both r_e and D_e values) for the ground states of these molecules. The success of this type of calculation *in these two cases* is impressive. Whether the method will be successful over a wide range of cases is still to be determined.

The V_2 Molecule. Experimental data for this molecule have recently become available[136]. This molecule closely resembles Cr_2 in having strong *d–d* bonding, with a bond length of 1.77 Å, a vibrational frequency of 535 cm^{-1} and a $^3\Sigma_g^-$ ground state, which can reasonably be attributed to an electron configuration of $\sigma^2\pi^4\sigma(s)^2\delta^2$. An SCF-HF-CI calculation[134)] on the V_2 molecule has given the $^3\Sigma_g^-$ ground state with $r_e = 1.77$ Å and a frequency of 594 cm^{-1}, in good agreement with experiment. It would appear that the 5-electron and 6-electron atoms, V and Cr, are both suited to giving strongly *d–d* bonded M_2 molecules, whereas all of the other *3 d* metal atoms preserve local d^n atomic configurations and form bonds only through the use of their *4s* orbitals.

The Ru_2 Molecule. In addition to the very thoroughly studied Mo_2 molecule, there is only one other second transition series molecule that has been shown from theoretical studies to have strong *d–d* bonding, although it has yet been observed experimentally. An SCF-HF-CI calculation on the Ru_2 molecule[137)] predicts a $^7\Delta_u$ ground state based on the following natural orbital populations: $(4\,d\sigma_g)^{1.67}(4\,d\pi_u)^{3.31}(4\,d\delta_g)^{2.73}(4\,d\delta_u)^{2.42}(4\,d\pi_g)^{2.60}$-$(4\,d\sigma_u)^{1.27}(5\,s\sigma_g)^{1.94}(5\,s\sigma_u)^{0.06}$. A bond length in the range 2.54–2.59 Å was also predicted.

5.2 Compounds With Chromium(II)-Chromium(II) Quadruple Bonds

The question of how these bonds behave as a function of the ligands by which they are surrounded has been the subject of a great deal of both experimental and theoretical study. Most of the molecular systems under consideration can be described by the general structural formula shown as 5.2-I, or some simple variant of it. The axial ligands, L

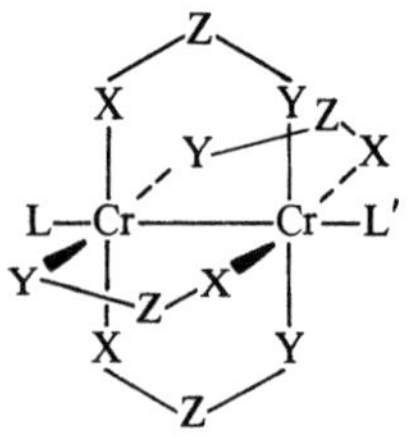

(5.2-I)

and L′, may be the same or different; one, or both, may be absent. The bridging ligands, XZY, are usually uninegative groups in which a single atom connects the donor atoms X and Y. Important examples are carboxyl anions, RCO_2^-, and amido anions, $RC(NR')O^-$, although still others such as $(CH_3)_2P(CH_2)_2^-$, $RC(NR')_2^-$ and those shown as 5.2-II and 5.2-III, have been used.

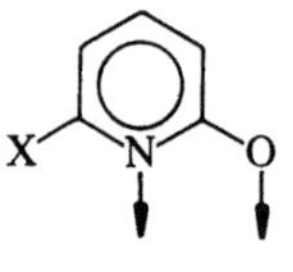

(5.2-II)

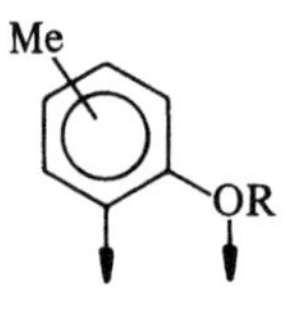

(5.2-III)

(a) The Experimental Position-I

One of the extraordinary characteristics of this series of molecules is the enormous variation in the Cr–Cr distance as the ligands are changed. The experimental data bear-

Table 5.2-1. Some carboxylato and amidato dichromium(II) compounds and their Cr–Cr bond lengths[a]

Compound	D(Cr–Cr), Å
1. $Cr_2(O_2CCH_3)_4(H_2O)_2$	2.362(1)
2. $Cr_2(O_2CCH_3)_4(py)_2$	2.369(2)
3. $Cr_2(O_2CCF_3)_4(Et_2O)_2$	2.541(1)
4. $Cr_2[O(RN)CCH_3]_4$[b]	1.873(7)
5. $Cr_2[O(R'N)CCH_3]_4$[c]	1.937(2)
6. $Cr_2[O(R'N)CCH_3]_4(THF)$	2.023(1)
7. $Cr_2[O(R'N)CCH_3]_4(THF)_2$	2.221(3)
8. $Cr_2[O(R'N)CCH_3]_4(py)_2$	2.354(5)

[a] Literature references for all these compounds will be found in Chap. 4 of MBBMA
[b] $R = C_6H_5$
[c] $R' = 2,6\text{-}Me_2C_6H_3$

Table 5.2-2. Some other dichromium(II) compounds and their Cr–Cr bond lengths[a]

Compound	D(Cr–Cr), Å
1. $Cr_2[Me_2P(CH_2)_2]_4$	1.895(3)
2. $Cr_2[(NMe)_2CPh]_4$	1.843(2)
3. $Cr_2(map)_4$[b]	1.870(3)
4. $Cr_2(mhp)_4$[c]	1.889(1)
5. $Cr_2(5\text{-}Me\text{–}2MeO\text{–}C_6H_3)_4$	1.828(2)
6. $Cr_2(O_2CCH_3)_2(2\text{-}Bu^tOC_6H_4)_2$	1.862(1)
7. $Cr_2(O_2CCH_3)_2[o\text{-}(NMe_2)C_6H_4CH_2]_2$	1.870(1)[d]
8. $Cr_2(fhp)_4(THF)$	2.150(2)[e]

[a] References for compounds 1–6 are given in Chap. 4 of MBBMA. For compounds 7 and 8 see Refs. given below
[b] map = anion of 6-methyl-2-amino-pyridine
[c] mhp = anion of 6-methyl-2-oxo-pyridine
[d] Cotton, F. A., Mott, G. N.: Organometallics, *1*, 302 (1982)
[e] Cotton, F. A., Falvello, L. R., Han, S., Wang, W.: Inorg. Chem. *22*, 4106 (1983)

ing on this point were reviewed in detail in MBBMA and the salient facts need be only briefly summarized here as a basis for discussing the work reported in the past 2–3 years. Tables 5.2-1 and 5.2-2 give lists of pertinent compounds and their Cr–Cr distances.

In Table 5.2-1 three representative carboxylato compounds (of the more than twenty that are structurally characterized) are listed. Each of them has two axial ligands and the bond lengths in these and all others are upwards of 2.30 Å. Compound 3 has about the longest Cr–Cr distance found in such compounds. From data such as these, and others available several years ago, it was inferred that the Cr–Cr distance was caused to increase by (a) stronger axial coordination, and (b) using carboxyl ligands from stronger acids. Recently, a very thorough study[138)] of a number of $Cr_2(O_2CR)_4L_2$ compounds, chosen to minimize steric effects, has provided further evidence that both of these factors are important in affecting those Cr–Cr bond lengths that are in the range of about 2.30 to about 2.55 Å.

Compounds 4–8 in Table 5.2-1 show another much more remarkable phenomenon. With amidato ligands, $CH_3CO(NR)^-$, in place of RCO_2^-, it is possible to obtain compounds without any axial coordination, and in these the Cr–Cr distances, ca. 1.9 Å, are *enormously* shorter than for *any* of the $Cr_2(O_2CR)_4L_2$ compounds. However, when axial ligands are added to $Cr_2[O(RN)CCH_3]_4$ molecules, the Cr–Cr bonds are lengthened in proportion to the extent of axial bonding. Thus, addition of one axial THF (compound 6) causes a increase of ca. 0.1 Å, while addition of two THF molecules causes an increase of ca. 0.3 Å (compound 7). In compound 8, where there are two axial pyridine ligands (stronger donors than THF), the bond has been lengthened even more, viz., by about 0.45 Å and is, in fact, essentially the same length as the one in the dipyridine adduct of the acetate. Clearly, when two axial pyridine ligands are present, the bond length is insensitive to whether the bridging ligands are RCO_2^- or $RCO(NR)^-$ ions.

It would seem difficult to escape the conclusion from these data that (a) while changes in both the bridging ligands and the axial ligands each have some influence on the Cr–Cr distance, the overwhelmingly more important factor is whether there are any axial ligands present, and how strongly they donate. When there are none present, at least for $Cr_2[O(RN)CR]_4$ molecules, the Cr–Cr bond lengths are only about 1.9 Å, or less. Bonds that are ≤ 1.9 Å have been dubbed "supershort", and we shall use that adjective in this sense in the following discussion.

Table 5.2-2 lists a number of other representative compounds with supershort Cr–Cr bonds. It is to be noted that they have a variety of combinations of atoms comprising the set of eight equatorial ligand atoms, viz., 8 C (in compound 1), 8 N (in compounds 2, 3), 4 N and 4 O (in compound 4); 4 C and 4 O (in compound 5), to deal only with those having four identical bridging ligands. There does not appear to be any particular pattern to the Cr–Cr distances in relation to the particular array of N, O or C atoms coordinated. Compound 6 in Table 5.2-2 is especially interesting because here, by using a mixed ligand set, two $CH_3CO_2^-$ ions and two alkoxyphenyl ligands, we have an equatorial ligand set made up of two C atoms and six O atoms. Still, the Cr–Cr distance is well down in the supershort range. Even in compound 7, where there are two ligands of somewhat different steric character together with two carboxyl groups, giving a 2 C + 2 N + 4 O ligand atom set, the distance is still supershort. The *one* thing that these seven compounds, as well as *all others with supershort bonds* have in common is: They have no axial ligands.

However, in compound 8, where we have a 4 O + 4 N ligand set (the fhp ligand is of type 5.2-II with X = F, and hence very similar to that in compound 4, where X = CH_3) the Cr–Cr distance is ca. 0.3 Å greater; this compound differs from the other seven in one obvious way: it has an axial ligand.

All of the foregoing observations very obviously lead to an interesting question: If one could measure the Cr–Cr distance in a $Cr_2(O_2CR)_4$ compound where there is no axial bonding, would it be within, or close to, the supershort range? On the basis of all of the observations summarized above, we think the answer to that question is *obviously* "yes". An experimental measurement of such a bond has proved very difficult to make since $Cr_2(O_2CR)_4$ compounds, which present no steric hindrance to axial coordination or molecular association through the axial positions (see MBBMA for details), have never yet provided a crystalline example of a compound with completely isolated $Cr_2(O_2CR)_4$ molecules. Nevertheless, based on the experimental data just discussed (all essential points of which were available in the literature several years ago) we think it is reasonable to expect the Cr–Cr bond in a $Cr_2(O_2CR)_4$ molecule to be ≤ 2.0 Å.

(b) Theoretical Work

There have been several theoretical treatments of the Cr–Cr quadruple bond[139–144], some of which have concerned themselves explicitly with what Cr–Cr distance would be expected for an isolated $Cr_2(O_2CR)_4$ molecule[142–144]. Molecules containing the Cr_2^{4+} core pose about as difficult a problem for *ab initio* calculations as does Cr_2, discussed earlier, and for the same reason: very strong electron correlation[139]. Indeed, correlation is so severe that the applicability of a one-electron-orbital formulation of the bonding has been viewed as questionable[140–141]. Since enormous amounts of configuration interaction have to be included in calculations on the dichromium complexes, some restraint has been necessary in the size of the basis sets. These difficulties in the calculations must be kept in mind when assessing the accuracy of the results.

Three independent studies[142–144] have attempted to determine how the "inductive effect of the bridging ligand" affects the equilibrium Cr–Cr bond length, all employing some form of SCF-HF-CI calculation. One of them[143] ventured no specific bond length predictions, but concluded that "variation in the electrostatic potential at the metal centre, due to the different ligands... leads to *large* variations in the Cr–Cr separation", and, further that "these separations are *mainly* determined by the electrostatic potential at the metal centre..." Italics in the above quotations have been added; the phrase "electrostatic potential at the metal centre" is another way of referring to the inductive effect of the bridging ligands.

Another of these studies[142] implied that for amidato-bridged species a short distance would be expected (this, of course, was already known to be true) but predicted "an equilibrium bond distance of 2.4 Å for tetrakis(formato)dichromium", and emphasized that "the nature of the bridging ligand *strongly* affects the strength of the quadruple bond". (Italics again added.)

Finally, the third paper[144] flatly predicted that the Cr–Cr distances in $Cr_2(NHCHO)_4$ and $Cr_2(O_2CH)_4$ should be 1.92 Å and 2.53 Å and stated that this large difference "shows that the effect of bridging ligands is at least as important as the observed influence of axial coordination in monitoring the length of the Cr$\overset{4}{-}$Cr bond". These authors went on to attribute to axial water molecules a capacity to further lengthen the bond in $Cr_2(O_2CH)_4$ by "0.05 Å only".

In summary, various theoretical groups have concluded from *ab initio* calculations that in systems of type I the Cr–Cr bond length is *strongly* influenced by changing the bridging ligands from those providing the more inductive sets of X and Y atoms (such as 8 C, 8 N, 4 N + 4 O) to an 8 O set, derived from four carboxyl groups. In several cases it has been explicitly predicted[142, 144] that while the Cr–Cr distance in $Cr_2[(NH)OCH]_4$ should be (as it was, of course, known to be) ca. 1.9 Å, it should be 2.4 Å or even >2.5 Å in $Cr_2(O_2CH)_4$.

(c) The Experimental Position-II

Although no way to obtain a crystalline compound containing a $Cr_2(O_2CR)_4$ molecule having no axial interactions has yet been found, there is at least one other possible way to obtain the desired bond length information, namely, by determination of the structure of such a molecule by electron diffraction in the gas phase. Only one such molecule has the necessary characteristics (sufficient vapor pressure at a temperature where decomposition is negligible, together with a sufficiently small and simple R group), namely the acetate. The formate cannot be vaporized without decomposition and $Cr_2(O_2CCF_3)_4$ has too many heavy atoms, as do other aliphatic carboxylates. Even so, the process of collecting and analyzing data for the acetate required extraordinary effort and very sophisticated methods of measurement and data analysis[145]. The result, however, is accurate and unambiguous: The Cr–Cr distance is 1.966(14) Å. This is completely in accord with what might have been expected empirically from experimental data on related compounds, but *completely at variance* with the predictions from ab initio calculations.

(d) Conclusions

Although the more rigorous, quantitative forms of molecular electronic structure theory are becoming steadily more powerful, and have certainly reached the point where they can provide valuable guidance to those doing experimental work on multiple bonds between metal atoms, things have not yet reached the point where the results can be taken uncritically. All such calculations are subject to error. In relatively simple molecules, and particularly molecules containing only light atoms, the probable magnitude of error can be reliably forecast, given the parameters of the calculation. Moreover, for smaller, lighter molecules it is computationally feasible to reduce the errors to almost any desired level by expanding the basis set and the amount of CI included. In a molecule such as $Cr_2(O_2CH)_4$, this is not yet the case. The inherent sources of error in the SCF-HF-CI method tend (predictably) to be at their worst for a case such as the Cr–Cr quadruple bond. However, at this time there is too little experience with such molecules to allow us to say *á priori* just *how* serious the errors will be.

It seems to us reasonable to believe that cautious empirical projection and "common sense" still have a dominant role to play in forecasting the behavior of molecules with multiple bonds between metal atoms and that when these approaches come into conflict with the results of *ab initio* calculations, the latter should be viewed with reservations.

We have already noted that as changes are made in the sets of atoms bonded equatorially to Cr_2^{4+} the actual observed Cr–Cr distances do not change very much. For example, for all sets, from the most "inductive" (8C; 8N; 4C + 4N) to intermediate ones (4N + 4O) to the nearly all oxygen set (2C + 6O) the Cr–Cr bond lengths are all within the range 1.82–1.93 Å, and there is no clear correlation of distance with ligand set. Another set of experimental facts to be noted is that $Cr_2[O(NR)CR']_4$ species *with* axial ligands behave enough like the $Cr_2(O_2CR)_4$ species with axial ligands, that it is reasonable to expect considerable similarity when the two kinds of molecules are compared *without* axial ligands. It would not seem to be in accord with a common sense interpretation of these facts to expect that merely changing from the 4N + 4O or, *á fortiori,* the 2C + 6O ligand atom sets to the 8O set found in $Cr_2(O_2CH)_4$ would cause the Cr–Cr distance would leap from the range of 1.82–1.93 Å out to 2.4 or 2.5 Å. Chemistry just doesn't work that way.

Thus, the experimental Cr–Cr bond length in $Cr_2(O_2CCH_3)_4$ is not in the least surprising from an empirical, common sense point of view. Instead, we need to understand better how to handle *ab initio* calculations on such molecules.

5.3 Platinum(III)-Platinum(III) Bonds

At the time MBBMA was written, little was known of the chemistry of compounds containing the Pt_2^{6+} central unit, which is isoelectronic with the Rh_2^{4+} unit. The scope of this area of platinum chemistry, and some that are closely related, has increased enormously in the past few years.

Actually, the very earliest reports of diplatinum(III) chemistry go back quite a long way. Apparently a French chemist, M. Blondel, was the first to describe such compounds, in 1905[146]. From a reaction of platinum(II) oxide with sulfuric and oxalic acids he obtained a product to which he assigned the formula "Pt_2O_3, 3 SO_3, SO_4H_2 + 11.5 H_2O". He also reported sodium and potassium salts of this acidic substance. He correctly decided on a dinuclear formula on the basis of his way of expressing the composition, i.e., as derived from an oxide. The compounds he made we now recognize as the $M^I_2[Pt_2(SO_4)_4(H_2O)_2] \cdot x H_2O$ (M^I = H, Na, K) compounds. A few years later Wöhler and Frey[147] and Delépine[148] confirmed Blondel's work and developed new methods to prepare these and related compounds.

Nothing further appears to have been published concerning these compounds until 1974 when a Russian team[149] reported anew the preparation of compounds containing the $[Pt_2(SO_4)_4(H_2O)_2]^{2-}$ ion, and in 1976 they provided the first unequivocal proof that a Pt–Pt bond is present in this type of binuclear anion[150]. Since then, there have been a number of further studies of the sulfato-bridged and related ortho phosphato-bridged diplatinum(III) species. In Table 5.3-1 are listed all the structurally characterized compounds and their Pt–Pt bond lengths.

The method of preparing the sulfato bridged diplatinum(III) species is based on heating a nitro complex of platinum(II), such as *cis*-$Pt(NH_3)_2(NO_2)_2$ or $[Pt(NO_2)_4]^{2-}$ in concentrated sulfuric acid. Oxides of nitrogen are evolved and the nature of the reaction is problematic. However, the product compounds can be obtained in good yields. The original structural work[150] on $K_2[Pt_2(SO_4)_4(H_2O)_2]$ was only incompletely described; the structure was later redone and fully reported[151]. The structure of the anion is depicted in Fig. 5.3-1. A compound in which the axial ligands are O-bonded dimethyl sulfoxide molecules, $K_2[Pt_2(SO_4)_4(OSMe_2)_2] \cdot 4 H_2O$, has also been studied crystallographically and it has an essentially similar structure.

The first report of orthophosphate-bridged species described a complex "$(NH_4)_2\{(H)_4[Pt_2(PO_4)_4(H_2O)_2\}$" and several derivatives thereof[152]. Further preparative and structural investigations have provided a detailed picture of the chemistry of these compounds. The sodium analog of the above compound was obtained in crystalline form

Table 5.3-1. Pt–Pt bond lengths in Pt_2^{6+} compounds. Sulfato and phosphate bridged compounds

Compound	Distance (Å)	Ref.
$K_2[Pt_2(SO_4)_4(H_2O)_2]$	2.461(1)	150, 151
$K_2[Pt_2(SO_4)_4(OSMe_2)_2] \cdot 4 H_2O$	2.471(1)	156
$Na_2[Pt_2(HPO_4)_4(H_2O)_2]$	2.486(1)	151, 153
$(pyH)[Pt_2(H_2PO_4)(HPO_4)_3(py)_2] \cdot H_2O$	2.494(1)	154
$(3,4\text{-}Me_2pyH)[Pt_2(HPO_4)_4(3,4\text{-}Me_2py)_2]$	2.494(1)	155
$[Et_4N]_2[Pt_2(H_2PO_4)_2(HPO_4)_2Cl_2] \cdot H_2O$	2.529(1)	151

and its structure determined[153] in 1982. It contains a $[Pt_2(HPO_4)_4(H_2O)_2]^{2-}$ anion, Fig. 5.3-2, quite analogous to the sulfator-bridged anions, with practically the same Pt–Pt distance (see Table 5.3-1). The preparative method involves heating *cis* or *trans*-$Pt(NH_3)_2(NO_2)_2$ in 84% H_3PO_4 to 100° or higher, followed by addition of suitable $M^I_2HPO_4$ salts and water to the cooled solution. The nature of the reaction is poorly understood, to say the least. There is evolution of NO gas, which indicates that NO_2^- is serving as the oxidizing agent for the Pt^{II}.

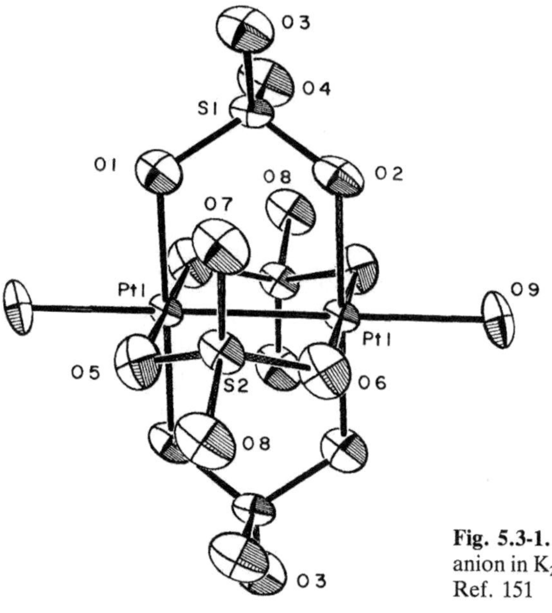

Fig. 5.3-1. The structure of the $[Pt_2(SO_4)_4(H_2O)_2]^{2-}$ anion in $K_2[Pt_2(SO_4)_4(H_2O)_2]$. By permission from Ref. 151

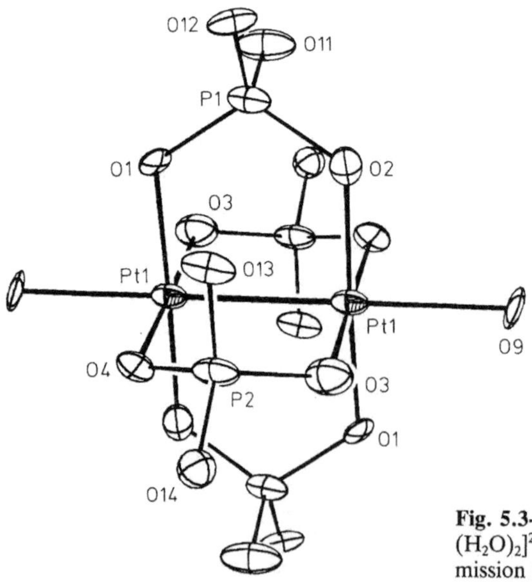

Fig. 5.3-2. The structure of the $[Pt_2(HPO_4)_4$-$(H_2O)_2]^{2-}$ anion in $Na_2[Pt_2(HPO_4)_4(H_2O)_2]$. By permission from Ref. 153

When pyridine or methylpyridines are added to the cooled reaction mixture, compounds containing such bases as axial ligands as well as the protonated bases as cations are obtained[154, 155]. However, the extent of protonation of the bridging orthophosphate ligands can vary. With 3,4-dimethylpyridine the straightforward product $[3,4\text{-}Me_2C_5H_3NH]_2[Pt_2(HPO_4)_4(3,4\text{-}Me_2C_5H_3N)_2]$ was obtained and structurally characterized[155]. However, with pyridine, a compound containing one $H_2PO_4^-$ and three HPO_4^{2-} ligands, $(pyH)[Pt_2(H_2PO_4)(HPO_4)_3(py)_2] \cdot H_2O$, was obtained and identified by X-ray crystallography[154]. Finally, along this same line, a compound containing two bridging $H_2PO_4^-$ ions and two bridging HPO_4^{2-} ions, $[NEt_4]_2[Pt_2(H_2PO_4)_2(HPO_4)_2Cl_2] \cdot H_2O$, has been characterized[151]. It is clear that compounds containing various mixtures of $H_2PO_4^-$ and HPO_4^{2-} ligands are all inherently stable and that their formation and isolability is determined by a combination of pH and solubility factors. However, systematic control of these factors has not yet been reported and the ligand combinations given above have been obtained by chance.

Let us turn now to other types of dinuclear compounds with Pt–Pt bonds that are somewhat longer than those in the compounds just discussed. These are listed in the lower part of Table 5.3-2.

In 1978, a diplatinum(III) compound with two bridging trifluoroacetate groups and four methyl groups in *cis* configurations about the platinum atoms was reported[157]. The $Pt_2(O_2CCF_3)_2(CH_3)_4(p\text{-}NC_6H_7)_2$ complex was prepared by oxidation of a platinum(II) species with silver(I). The reported Pt–Pt bond length (2.557(1) Å) is significantly longer than those in the sulfate and phosphate complexes. This is probably due to the presence of only two $CF_3CO_2^-$ ligands.

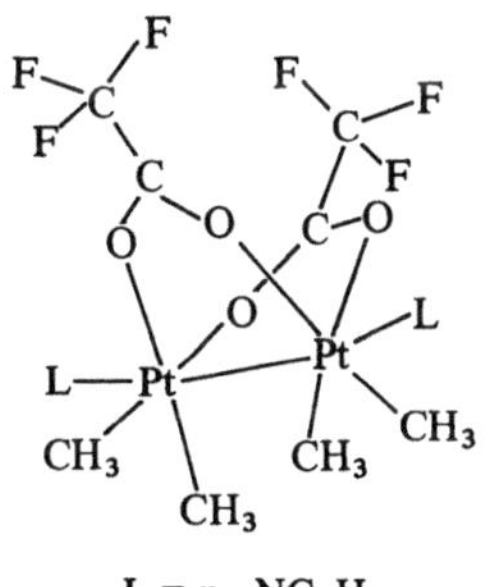

L = p–NC$_6$H$_7$

Table 5.3-2. Pt–Pt bond lengths in other Pt_2^{6+} systems

Compound	Distance (Å)	Ref.
$[(H_2O)(NH_3)_2Pt(C_5H_4NO)_2Pt(NH_3)_2(NO_3)](NO_3)_3 \cdot 2\ H_2O$	2.539(1)	159
$[(NO_3)(NH_3)_2Pt(C_5H_4NO)_2Pt(NH_3)_2(NO_3)](NO_3)_2 \cdot H_2O$	2.547(1)	160
$Pt_2(O_2CCF_3)_2(CH_3)_4(p\text{-}CH_3C_5H_4N)_2$	2.557(1)	157
$[Pt_2(NH_3)_4(C_5H_4NO)_2Cl_2](NO_3)_2$	2.5684(5)	161
$[(NO_3)(NH_3)_2Pt(C_5H_4NO)_2Pt(NH_3)_2(NO_2)](NO_3)_2 \cdot H_2O$	2.575(1)	160
$[Pt_2(NH_3)_4(C_5H_4NO)_2Br_2](NO_3)_2 \cdot 0.5\ H_2O$	2.5820(6)	161
$K_4[Pt_2(P_2O_5H_2)_4Cl_2]$	2.695(1)	162

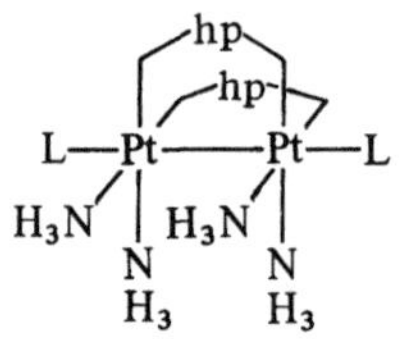

L = H_2O, NO_3^-, NO_2^-, Cl^-, Br^-

Closely related to $Pt_2(O_2CCF_3)_2(CH_3)_4(p\text{-}NC_6H_7)_2$ are the 2-oxopyridine (hp) compounds of diplatinum(III). These compounds are formed by oxidative cleavage of the 2-oxopyridine platinum blue tetramer[158]. The structures have the two hp ligands in a *cis*-orientation, four *cis*-amino groups, and a variety of axial ligands coordinating the central metal dimer unit. The Pt–Pt bond lengths are listed in Table 5.3-2. These bond lengths are close to the metal-metal separation in the complex $Pt_2(O_2CCF_3)_2(CH_3)_4(p\text{-}NC_6H_7)_2$ which also has two bridging ligands and are thus longer than the distances seen in the tetra-bridged sulfate and phosphate compounds.

In 1982, novel diplatinum(II) diphosphite complexes were reported[162]. These compounds were readily generated by oxidative addition to the binuclear platinum(II) tetrakis(diphosphite) complex, $[Pt_2(pop)_4]^{4-}$ (pop = $P_2O_5H_2^{2-}$), with halogens or methyl

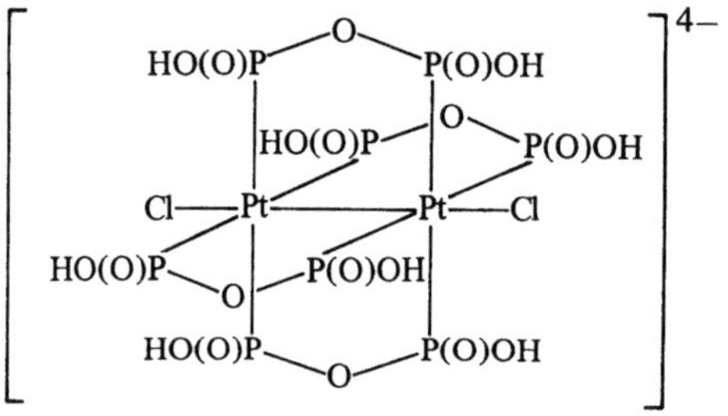

iodide. The $[Pt_2(pop)_4Cl_2]^{4-}$ ion reported has a Pt–Pt bond length of 2.695(1) Å, which is the longest platinum(III)-platinum(III) bond known. It was stated by the authors that oxidative addition to binuclear Pt(II) was favorable energetically and indeed this method may be a productive synthetic pathway to platinum(III) dimers.

5.4 Chemistry of Dimolybdenum and Ditungsten Hexalkoxides

The chemistry of the M_2X_6 molecules (M = Mo, W and X = R, OR, NR_2, etc.) has continued to develop rapidly, with many of the major advances coming from the laboratory of M. H. Chisholm. There is, in fact, far too much new material to be reviewed comprehensively here. We shall restrict coverage to some of the reaction chemistry of the $M_2(OR)_6$ molecules, where particularly important advances have been made, namely, to the reactions of the $M_2(OR)_6$ molecules with C≡C and C≡N bonds.

With alkynes, R′C≡CR″, the $M_2(OR)_6$ molecules give a wide variety of products, depending on (a) conditions, (b) the R′ and R″ in the alkyne, (c) the OR group in the

Table 5.4-1. Selected bond distances (Å) for $M_2(OR)_6(py)_n(\mu\text{-}C_2R'_2)$ compounds of molybdenum and tungsten (n = 1 or 2)

Compound	Structure type[a]	M–M	C–C (averaged)	M–C	Ref.
$Mo_2(O\text{–}i\text{-}Pr)_6(py)_2(\mu\text{-}C_2H_2)$	I	2.554(1)	1.37(1)	2.09	b
$W_2(O\text{–}i\text{-}Pr)_6(py)_2(\mu\text{-}C_2H_2)$	I	2.567(1)	1.39(2)	2.09	c
$W_2(OCH_2\text{–}t\text{-}Bu)_6(py)_2(\mu\text{-}C_2H_2)$	II	2.610(2)	1.39(3)	2.14	d
$W_2(OCH_2\text{–}t\text{-}Bu)_6(py)_2(\mu\text{-}C_2Me_2)$	II	2.602(1)	1.37(2)	2.14	c
$W_2(OCH_2\text{–}t\text{-}Bu)_6(py)_2(\mu\text{-}C_2Et_2)$	III	2.572(1)	1.40(2)	2.14	d
$W_2(O\text{–}t\text{-}Bu)_6(py)(\mu\text{-}C_2H_2)$	III	2.665(1)	1.44(1)	2.10	c

[a] I, II and III refer to the schematic drawings given in Fig. 5.4-1
[b] Chisholm, M. H., Folting, K., Huffman, J. C., Rothwell, I. P.: J. Am. Chem. Soc. *104,* 4839 (1982)
[c] Chisholm, M. H., Folting, K., Hoffman, D. M., Huffman, J. C.: J. Am. Chem. Soc. *106,* 6794 (1984)
[d] Chisholm, M. H., Conroy, B. K., Folting, K., Hoffman, D. M., Huffman, J. C.: results to be published

metal compound and (d) the identity of the metal atoms (i.e., Mo or W). The question of conditions refers primarily to temperature, to whether a good donor ligand, usually pyridine, is present or not, and to the mole ratio of alkyne to $M_2(OR)_6$. Generally speaking the presence of pyridine, which becomes incorporated into many of the products, is advantageous in giving smoother, cleaner reactions.

Simple crosswise addition of the R′C≡CR″ molecule to the M≡M bond of the $M_2(OR)_6$ molecule, in the presence of pyridine, has been observed in many cases and the principal results are summarized in Table 5.4-1 (we are grateful to M. H. Chisholm for providing much of this information in advance of publication). Drawings showing the three structure types are presented in Fig. 5.4-1. The factors determining which of these structures is adopted by a given product are not fully understood. Only molybdenum has given structures of type I. The tendency of alkynes to bind more tightly to the ditungsten center may account for the adoption by the tungsten compounds of structures II and III. Bulky R, R′ and R″ groups may also favor III. In addition, however, the structural question has another aspect since in solution several of these molecules have been shown by NMR to be fluxional.

At a higher stage of complexity are alkyne coupling reactions. All of the isolable simple alkyne addition products are reactive toward further alkyne molecules to give new products in which two or more alkyne molecules have been coupled[163–165]. Two of these are shown in Fig. 5.4-2. In the case of the molybdenum alkyne adducts of type I as well as the molybdenum compound shown in Fig. 5.4-2, it has been found[163] that they function catalytically to promote cyclotrimerization of alkynes to give benzenes. By isotopic labelling studies it has been shown that these coupling reactions are true addition reactions without any C–C bond breaking.

Alkyne scission reactions also occur and some are of exceptional interest and importance. Schrock, et al.[166] discovered the following remarkable reaction:

$$W_2(OCMe_3)_6 + RC{\equiv}CR \rightarrow 2\ (Me_3CO)_3W{\equiv}CR$$

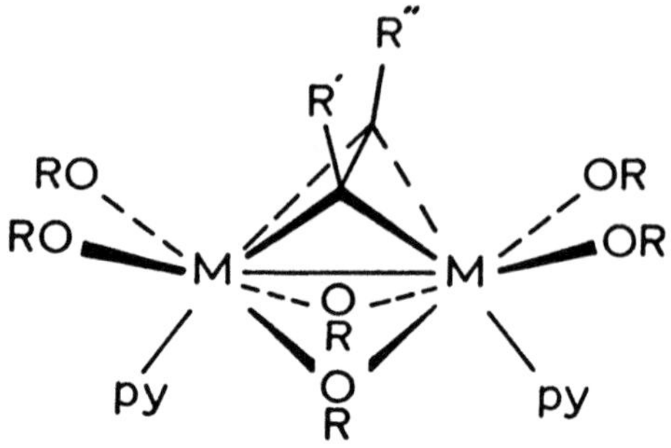

M = Mo or W R = i-Pr R' = R" = H
M = Mo R = i-Pr R' = H R" = Me; R' = R" = Me
M = Mo R = CH_2-t-Bu R' = R" = H

I

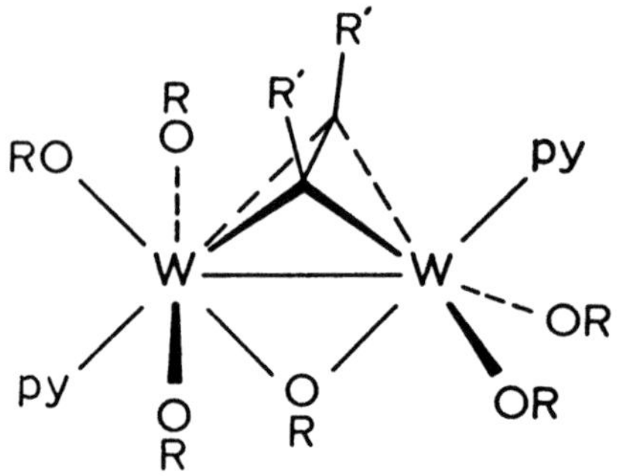

R = CH_2-t-Bu R' = H or Me

II

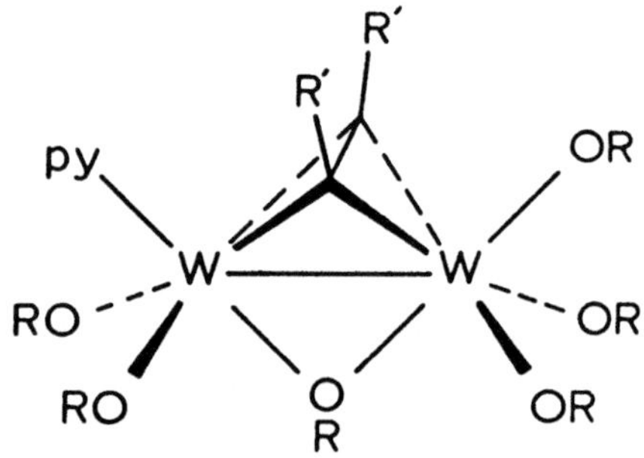

R = t-Bu R' = H
R = CH_2-t-Bu R' = Et or Ph

III

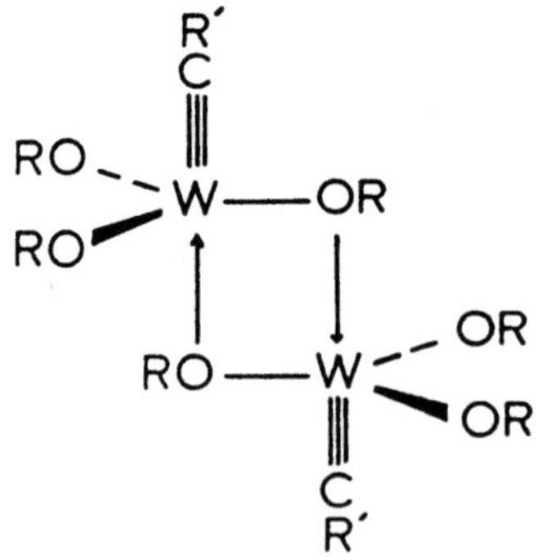

R = t-Bu
R' = Me or NMe_2

IV

Fig. 5.4-1. I, II and III: three structural types of alkyne adducts of $M_2(OR)_6$ and $W_2(OR)_6$ molecules. IV: The mode of dimerization of two $(Me_3CO)_3W{\equiv}CR$ compounds

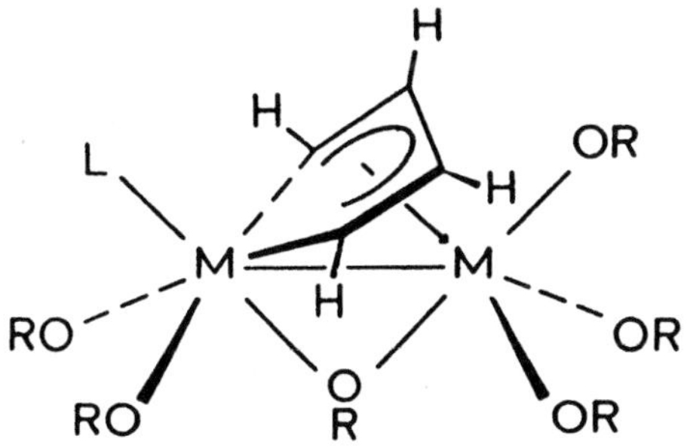

M = Mo or W	R = CH_2-t-Bu	L = py
M = W	R = t-Bu	L = none
M = W	R = CH_2-t-Bu	L = PMe_3
M = W	R = i-Pr	L = PMe_3

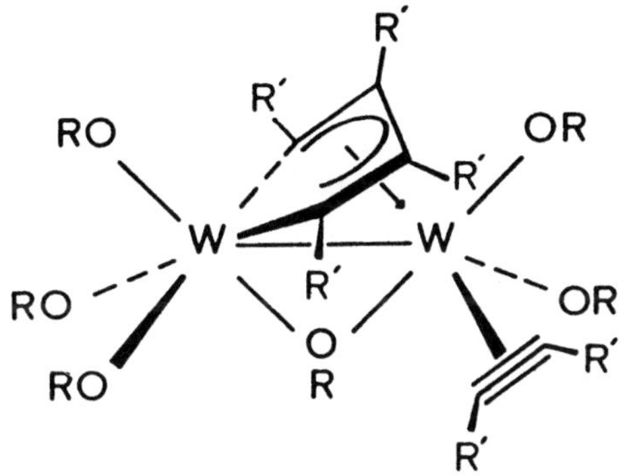

R = i-Pr	R' = H or Me
R = CH_2-t-Bu	R' = H or Me

Fig. 5.4-2. Two types of structures for products of alkyne coupling reactions

It is important to note that this reaction works only with $W_2(OCMe_3)_6$ and not with any other $M_2(OR)_6$ molecule and that while it appears to be general for aliphatic R it does not proceed for R = C_6H_5. It goes under very mild conditions and is virtually quantitative. Since the $(Me_3CO)_3W{\equiv}CR$ compounds are alkyne metathesis catalysts, still other tungsten alkylidyne compounds, not obtainable in the above reaction, can be obtained by the following reaction:

$$W_2(OCMe_3)_6 + 2\ EtC{\equiv}CR \rightarrow 2\ (Me_3CO)_3W{\equiv}CR + EtC{\equiv}CEt$$

[R = Ph, $SiMe_3$, CH_2=CH]

The $(Me_3CO)_3W{\equiv}CR$ compounds exist in the solid state as either weakly associated dimers (Structure IV, Fig. 5.4-1) with OR bridges (R = CH_3[167] or NMe_2)[168] or as a tetrahedral monomer in the case of R = C_6H_5[169].

Under more severe conditions $W_2(OCMe_3)_6$ reacts directly with both PhC≡CPh and EtC≡CEt to give a variety of more complex products[170], such as $W_2(OCMe_3)_4(\mu\text{-CPh})_2$, $W_2(OCMe_3)_4(PhCCPh)_2$, and $[W_3(OCMe_3)_5(\mu\text{-O})(\mu\text{-CEt})O]_2$. $(Me_3CO)_3W{\equiv}CR$ does not appear to be an intermediate in these reactions. On the other hand, $W_2(OCHMe_2)_6(py)_2$ reacts with one equivalent of MeC≡CMe to give, via an $(RO)_3W{\equiv}CMe$ intermediate, a triangulo cluster with a capping alkylidyne ligand, $W_3(\mu_3\text{-CMe})(\mu\text{-OCHMe}_2)_3(OCHMe_2)_6$[171]. It is believed that this product arises by a two-step process in which $(Me_2CHO)_3W{\equiv}CMe$ is first formed but reacts rapidly with a molecule of $W_2(OCHMe_2)_6$

to give the trinuclear product. The bulkier $OCMe_3$ groups presumably impede this process, thus allowing only the $(Me_3CO)_3W{\equiv}CR$ molecules to be formed.

$W_2(OCMe_3)_6$ reacts with nitriles in a way similar to its reactions with alkynes, as indicated in the following general reaction[166]:

$$W_2(OCMe_3)_6 + RC{\equiv}N \rightarrow (Me_3CO)_3W{\equiv}N + (Me_3CO)_3W{\equiv}CR$$

The nitrido compound is very stable and only slightly soluble and thus this reaction provides access to some alkylidyne compounds not readily accessible by the reaction with alkynes, e.g., $(Me_3CO)_3W{\equiv}CPh$. The solid $(Me_3CO)_3W{\equiv}N$ has a linear polymeric structure[167] in which the ··· WNWN ··· chains have alternating W–N distances of 1.74(1) Å and 2.66(1) Å.

Other $W_2(OR)_6$ compounds (R = $CHMe_2$ and CH_2CMe_3) also react with nitriles affording the $(RO)_3W{\equiv}N$ compounds. Presumably $(RO)_3W{\equiv}CR'$ molecules are also formed but in no case have they seen isolated. It is interesting that no similar reaction between a nitrile and a $Mo_2(OR)_6$ compound has yet been found. The $Mo_2(OR)_6$ compounds, in fact, appear to be entirely inert towards nitriles and do not even form simple adducts with them.

5.5 *Photoelectron Spectroscopy*

This technique has continued to be used in the conventional manner to study compounds with M–M multiple bonds and has provided useful information on the electronic structure of several molecules. An important example of the conventional, but thorough, use of the technique is provided by a study of the $Mo_2Cl_4(PMe_3)_4$ and $W_2Cl_4(PMe_3)_4$ molecules[93]. The spectra of both molecules were obtained using both HeI and HeII excitation, as shown for the tungsten compound in Fig. 5.5-1. It is clear that the comparison of the two measurements provides unambiguous evidence for assigning the ionizations that arise from heavily metal-based orbitals, namely the first two at ca. 6 eV and ca. 7.2 eV (doublet). The doublet structure of the second ionization was shown to be of the expected form for a π orbital which is split by spin-orbit coupling of one component to the δ level and the other to the σ level. The possibility that the σ ionization is also contained under the envelope of this doublet is effectively excluded. When the δ and π ionization energies for the molybdenum and tungsten molecules are compared, it is seen that they are appreciably lower for the tungsten compound and this provides an explanation for characteristic chemical differences between the quadruply-bonded dimolybdenum compounds as a class and the quadruply-bonded ditungsten compounds as a class. Some of the consequences of these differences have already been discussed in detail in Sec. 3 of this article.

Of even greater interest, perhaps, because of its potential for delving still more deeply into the bonding in these molecules is the high resolution PES technique developed by Lichtenberger[172]. In a study of $Mo_2(O_2CCH_3)_4$, it was possible to resolve and analyze a vibrational progression on the δ ionization band. The band contour and the explanation in terms of the ν(Mo–Mo) vibration of the $^2B_{1g}$ state of the $[Mo_2(O_2CCH_3)_4]^+$ ion are shown in Fig. 5.5-2. In their interpretation of these results Lichtenberger and Blevins were able to provide virtually conclusive support for assignment of this particular ioniza-

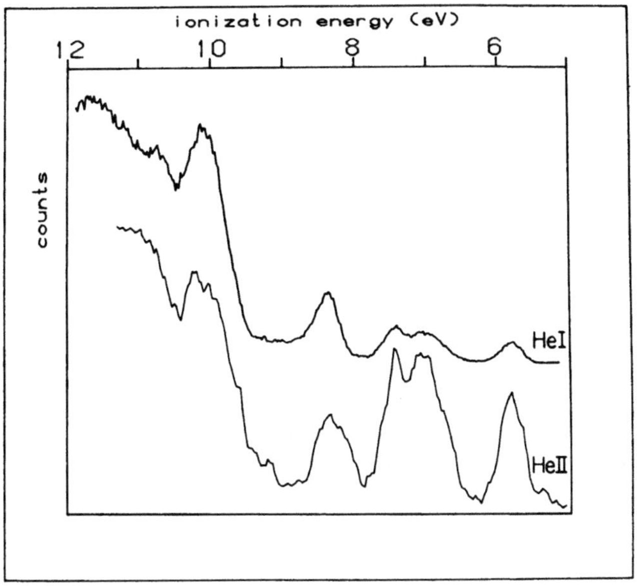

Fig. 5.5-1. The photoelectron spectrum of the $W_2Cl_4(PMe_3)_4$ molecule in the gas phase. By permission from Ref. 93

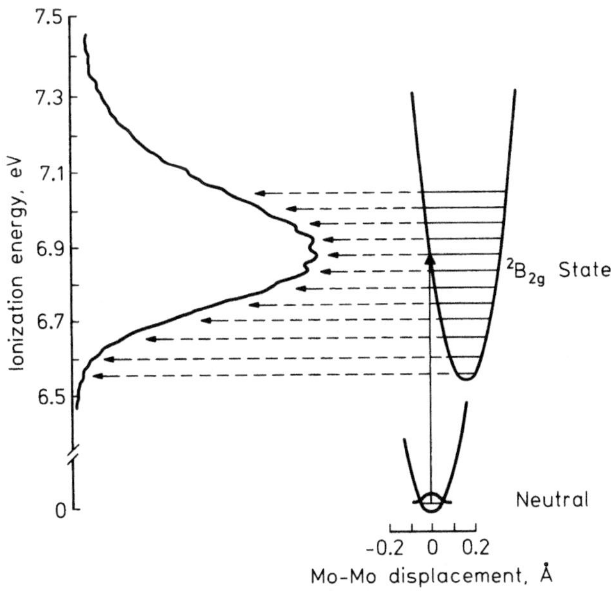

Fig. 5.5-2. The Franck-Condon analysis of the δ ionization peak for $Mo_2(O_2CCH_3)_4$, showing how the Mo–Mo bond length change is estimated. By permission from Ref. 172

tion to the δ orbital and from an interpretation of the Franck-Condon factors of the vibrational progression they were able to estimate that in the ion the Mo–Mo distance must be 0.18 ± 0.02 Å longer than in the $^1A_{1g}$ ground state of the $Mo_2(O_2CCH_3)_4$ molecule. This, however, is a rather large change and cannot be understood purely in terms of a bond order reduction of only 0.5 unit of δ bonding. The proper interpretation (which was given by Lichtenberger and Blevins) can best be appreciated within the context of other work bearing on the same problem. This will be the subject of the next subsection, 5.6.

5.6 M–M Bond Lengths vs. Bond Order and Charge

In dealing with multiple bonds between metal atoms, especially the δ components of such bonds, it has been found that the bond length and bond order are not necessarily inversely related[173)], even though such a rule had previously been thought to apply universally and no exceptions were previously known. The first exception was provided by a comparison of the structures of the $[Tc_2Cl_8]^{3-}$ and $[Tc_2Cl_8]^{2-}$ ions, in which the formal bond orders are 3.5 ($\sigma^2\pi^4\delta^2\delta^*$) and 4.0 ($\sigma^2\pi^4\sigma^2$) whereas the Tc–Tc distances are 2.11 ± 0.005 Å and 2.15 Å, respectively[14)]. The following explanation was proposed for this. In going from the 3− to the 2− ion, the mean oxidation number of the metal atoms is increased from +2.5 to +3.0. This causes all the metal *d* orbitals to contract and thus weakens all component of the Tc–Tc bonding, including the dominant σ and π components. This bond-weakening, and hence bond-lengthening, effect more than offsets the bond-strengthening effect of removing a δ* electron and the net result is an increase in the bond length.

A number of other lines of evidence have now been developed to show that the above argument is correct. In the series of isostructural species[48)] shown below the Re–Re distances have been found to vary rather little even though the bond order changes from

$[Re_2Cl_4(PMe_2Ph)_4]^0$	2.241(1) Å
$[Re_2Cl_4(PMe_2Ph)_4]^+$	2.218(1) Å
$[Re_2Cl_4(PMe_2Ph)_4]^{2+}$	2.215(2) Å

3.0 to 3.5 to 4.0. Through this series we have two countervailing trends: (1) because of the increasing oxidation number of the metal atom, the bonds are tending to get longer; (2) because of the increasing bond order (from a $\sigma^2\pi^4\delta^2\delta^{*2}$ to a $\sigma^2\pi^4\delta^2$ configuration) the bonds are tending to get shorter. In the first step the bond order effect is slightly more important while in the second step the two factors are almost exactly balanced.

In the case of the $[Mo_2(O_2CCH_3)_4]^0$ to $[Mo_2(O_2CCH_3)_4]^+$ ionization, discussed in Sect. 5.5, the rather large change in bond length inferred from the Franck-Condon factors is clearly attributable to the fact that both the bond order and oxidation number effects are acting in the same direction.

It is obvious that the validity of any hypothesis in which two factors are postulated to be operating simultaneously can be critically tested if cases can be found in which only one of them comes into play. Its effect can then be observed unambiguously. Thus, one question we are led to ask is whether it is possible to change the δ bond order, and measure the accompanying change in bond length, without changing the ionization state of the metal atoms (or any other significant factor). While this might seem unlikely to be possible, there are actually two ways to do it, and both have been employed.

Just as the analysis of the Frank-Condon progression in the ionization of $Mo_2(O_2CCH_3)_4$ was used to estimate the increase in bond length resulting from that process, so the progressions observed in the $\delta \rightarrow \delta^*$ transitions can be similarly analyzed[173]. For $Mo_2(O_2CCH_3)_4$ and $[Mo_2Cl_8]^{4-}$ the results are about 0.11 Å and for several Re_2^{6+} species the change is about 0.07 Å.

A second way to change the δ bond order without altering the composition of the system in any way is to vary the angle of internal rotation, χ, over as wide a range as possible between the limits of eclipsed ($\chi = 0$) and staggered ($\chi = 45°$). This should cause the δ bond order to change from 1.0 to zero and if the bond order follows the overlap the functional dependence of bond order on χ will be according to $\cos 2\chi$. This idea was given a preliminary test using a set of three compounds[174] of the type $Mo_2X_4(LL)_2$ in which LL is a bridging bidentate ligand (e.g., a diphosphine) that causes some degree of twist away from the eclipsed conformation in order to accommodate the conformation requirements of the $\overline{\text{Mo–Mo–P–C}_n\text{–P}}$ rings. A very good straight line was obtained for a plot of Mo–Mo distance vs. $\cos 2\chi$. Subsequently, data were accumulated for ten such compounds[175] and essentially the same line was defined, with a correlation coefficient of 0.92. Over the entire range of $\cos 2\chi$ from 0 to 1, the Mo–Mo distance changes by 0.06 Å. This number is somewhat smaller than that (0.11 Å) given by the spectroscopic data. However, it seems that from a variety of data, the loss of the δ bond may be expected to increase the Mo–Mo distance by ca. 0.10 Å. This suggests that more than half of the change (of 0.18 Å) that occurs on ionization of $Mo_2(O_2CCH_3)_4$ may be due to the reduction of σ and π overlap.

While there is no direct evidence for the effect of a change in metal atom charge, with constant formal bond order, on the bond length among compounds with delta bonds, comparison of the M–M distances in several confacial bioctahedral species, $[M_2X_9]^{n-}$ indicate the importance of the effect[88]. For example, the W–W distance in $[W_2Cl_9]^{2-}$ is 0.12 Å longer than that in $[W_2Cl_9]^{3-}$ even though the electron being lost from the latter is expected on theoretical grounds to be virtually non-bonding. Also, when the strictly isoelectronic species $[W_2Cl_9]^{3-}$ and $[Re_2Cl_9]^{-}$ are compared, the increase of about 0.30 Å upon changing from two + 3 metal atoms to two + 4 metal atoms clearly shows how much an increase in charge can weaken the M–M bond.

5.7 *Optical Activity*

Many metal-metal multiple bonds have rich visible absorption spectra and the possibility of placing them in chiral environments and observing CD and/or ORD effects is an attractive one for several reasons. From one point of view, the optical effects can provide valuable information on the nature of the electronic transitions, thus helping with assignments. Conversely, for a sufficiently well understood electronic transition, the sign of the CD (or ORD) effect can provide information on the absolute configuration of the chiral molecule (or chiral ligands therein). Studies based on both of these considerations have

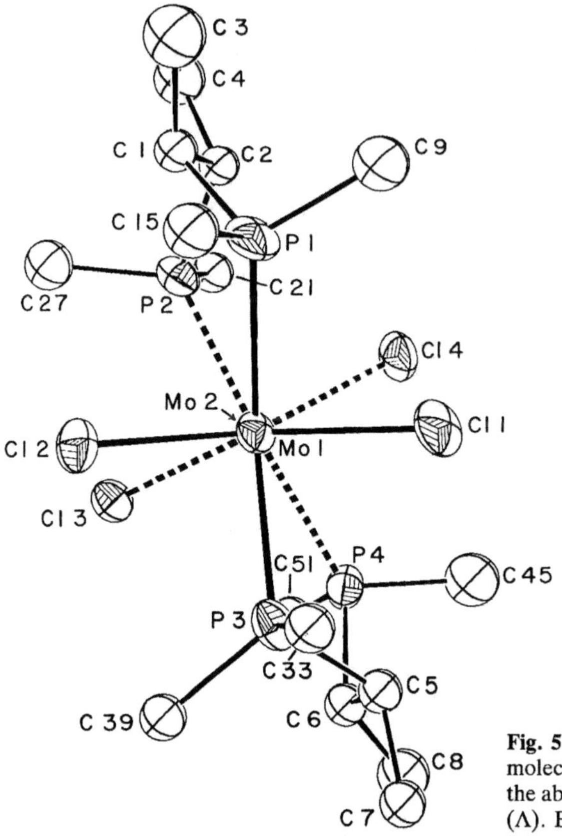

Fig. 5.7-1. A view of the Mo_2Cl_4(S,S-dppb)$_2$ molecule down the Mo–Mo axis showing that the absolute configuration is counterclockwise (Λ). By permission from Ref. 179

been published; although the work is still very preliminary, it demonstrates feasibility and a great deal more will doubtless be forthcoming.

It has been known for several years[176–178] that molecules of the type M_2X_4(diphosphinoethane)$_2$, in which the diphosphinoethane ligands bridge the M–M bond (the β isomers), can be prepared and that the conformational demands of the M–P–C–C–P rings cause a twisting about the M–M bond. These helical molecules are chiral, with either a right- or lefthand twist, but the enantiomers form in equimolar amounts if the diphosphine ligands themselves are achiral. However, by using a chiral phosphine such as S,S-$Ph_2PCH(CH_3)CH(CH_3)PPh_2$, dimetal molecules of only one chirality can be obtained.

In these β-M_2X_4(diphos)$_2$ type molecules, the electronic transitions that are mainly localized in the M–M bond should be strongly affected by the inherent chirality (helicity) of the inner $M_2X_4P_4$ group. When M = Mo or W, that is, for quadruple bonds, the well-known $\delta \rightarrow \delta^*$ transition is an especially attractive one to examine because the assignment is unequivocal and it is possible to establish an unambiguous, qualitative relationship between the absolute conformation of the helix and the sign of the CD effect[179].

In Fig. 5.7-1 is shown a view of the Mo_2Cl_4(S,S-dppb)$_2$ molecule (where S,S-dppb means S,S-(2,3-diphenylphosphino)butane) down the Mo–Mo axis. The absolute configuration is Λ (counter-clockwise) and a measurement of the CD spectrum showed a symmetrical negative CD feature centered at 13,700 cm^{-1}, which is the center of the $\delta \rightarrow \delta^*$ absorption band. Since the $\delta \rightarrow \delta^*$ transition is electric-dipole allowed with polarization along the molecular axis (the axis of the helix) it allows a particularly simple and unambiguous opportunity for deducing the relationship between the directions of the charge transfer vector and the charge rotation vector. If these two point in the same direction a positive CD effect is predicted, while if the vectors are oppositely directed the CD effect should be negative. It is easily shown (as explained in Ref. 179) that for a counterclockwise (Λ) twist the two vectors must point in opposite directions and thus the observed negative CD effect for the $\delta \rightarrow \delta^*$ transition is unambiguously predictable.

The work just discussed deals with the situation in which the chromophore *per se* is chiral. This is one of two main classes of chiral molecules recognized by specialists in the field[180, 181]; it is commonly called the case of "the inherently chiral chromophore". The second class of chiral molecules are those which have an "achiral chromophore in a chiral environment".

We turn now to the second type of study that is possible with compounds containing chiral ligands coordinated to M–M multiple bonds. In this case we are dealing with the "achiral chromophore (the M_2 unit) in a chiral environment". The published work is due to Snatzke[182–185] who has shown that the introduction of small amounts of $Mo_2(O_2CCH_3)_4$, or occasionally other M_2^{n+}-containing species, into solutions of chiral organic compounds allows one to observe CD and ORD effects in the visible region where measurements are easily made but where the organic compounds themselves do not normally have absorption bands. Snatzke has been able to develop empirically-based rules that allow the determination of absolute configurations. However, a fundamental, first-principles understanding of how the CD and ORD effects are imposed on the electronic transitions of the central M_2^{n+} units still has to be developed. Some work in this direction has been undertaken[186–188] and will be published in the near future.

6 References

1. Cotton, F. A., Walton, R. A.: Multiple Bonds Between Metal Atoms, J. Wiley & Sons, New York, 1982 and references therein
2. Chisholm, M. H., Rothwell, I. P.: Prog. Inorg. Chem. *29,* 1 (1982)
3. Felthouse, T. R.: Prog. Inorg. Chem. *29,* 73 (1982)
4. Chisholm, M. H. (Editor): ACS Symp. Series, No. 221, 1983; See articles by Lichtenberger, Clark and Stead, Curtis, Chisholm, Cotton, Walton and Sattelberger
5. Chisholm, M. H.: Polyhedron *2,* 681 (1983)
6. Cotton, F. A., Chisholm, M. H.: Chem. & Eng. News *60,* June 28, page 40 (1982)
7. Walton, R. A.: Israeli J. Chem., in press
8. Koz'min, P. A., Surazhskaya, M. D.: Sov. J. Coord. Chem. *6,* 309 (1980)
9. Preetz, W., Peters, G.: Z. Naturforsch. *35b,* 797 (1980)
10. Cotton, F. A., Daniels, L., Davison, A., Orvig, C.: Inorg. Chem. *20,* 3051 (1981)
11. Cotton, F. A., Pedersen, E.: ibid. *14,* 383 (1975)
12. Bratton, W. K., Cotton, F. A.: ibid. *87,* 921 (1965)
13. Cotton, F. A., Shive, L. W.: ibid. *14,* 2032 (1975)
14. Cotton, F. A., Davison, A., Day, V. W., Fredrich, M. F., Orvig, C., Swanson, R.: ibid. *21,* 1211 (1982)

15. Cotton, F. A., Gage, L. D.: Nouv. J. Chim. *1,* 441 (1977)
16. Zaitseva, L. L., Kotel'nikova, A. S., Rezvov, A. A.: Russ. J. Inorg. Chem. *25,* 1449 (1980)
17. Spitsyn, V. I., Baierl, B., Kryuchkov, S. V., Kuzina, A. F., Varen, M.: Dokl. Akad. Nauk SSSR *256,* 608 (1981)
18. Kryuchkov, S. V., Kuzina, A. F., Spitsyn, V. I.: ibid. *266,* 127 (1982)
19. Rakitin, Yu. V., Kryuchkov, S. V., Aleksandrov, A. I., Kuzina, A. F., Nemtsev, N. V., Ershov, B. G., Spitsyn, V. I.: ibid. *269,* 1123 (1983)
20. Koz'min, P. A., Larina, T. B., Surazhskaya, M. D.: Sov. J. Coord. Chem. *8,* 451 (1982)
21. Koz'min, P. A., Larina, T. B., Surazhskaya, M. D.: Koord. Khim. *7,* 1719 (1981) (Chem. Abs. *96,* 13966b (1982))
22. Cotton, F. A., Fanwick, P. E., Gage, L. D.: J. Am. Chem. Soc. *102,* 1570 (1980)
23. Nefedov, V. I., Koz'min, P. A.: Inorg. Chim. Acta *64,* L177 (1982)
24. Spitsyn, V. I., Kuzina, A. F., Kryuchkov, S. V.: Russ. J. Inorg. Chem. *25,* 406 (1980)
25. Kryuchkov, S. V., Kuzina, A. F., Spitsyn, V. I.: ibid. *28,* 1124 (1983)
26. Barder, T. J., Walton, R. A.: Inorg. Chem. *21,* 2510 (1982)
27. Bursten, B. E., Cotton, F. A., Fanwick, P. E., Stanley, G. G.: J. Am. Chem. Soc. *105,* 3082 (1983)
28. Hay, P. J.: ibid. *104,* 7007 (1982)
29. Clark, R. J. H., Stead, M. J.: Inorg. Chem. *22,* 1214 (1983)
30. Nocera, D. G., Gray, H. B.: J. Am. Chem. Soc. *103,* 7349 (1981)
31. Nocera, D. G., Maverick, A. W., Winkler, J. R., Che, C., Gray, H. B.: ACS Symp. Ser. *No. 211,* 21 (1983)
32. Koz'min, P. A., Surazhskaya, M. D., Larina, T. B.: Koord. Khim. *8,* 709 (1982) (Chem. Abs. *97,* 136870u (1982))
33. Calderazzo, F., Marchetti, F., Poli, R., Vitali, D., Zanazzi, P. F.: J. Chem. Soc., Dalton Trans. 1665 (1982)
34. Shtemenko, A. V., Bagirov, Sh. A., Kotel'nikova, A. S., Lebedev, V. G., Kazymov, O. I., Alieva, A. I.: Russ. J. Inorg. Chem. *26,* 58 (1981)
35. Misailova, T. V., Kotel'nikova, A. S., Golovaneva, I. F., Evstaf'eva, O. N., Lebedev, V. G.: ibid. *26,* 343 (1981)
36. Koz'min, P. A., Surazhskaya, M. D., Larina, T. B., Shtemenko, A. V., Kotel'nikova, A. S., Golovaneva, I. F.: Sov. J. Coord. Chem. *5,* 386 (1981)
37. Kotel'nikova, A. S., Moskovkin, A. S., Misailova, T. V., Miroshnichenko, I. V., Bagirov, Sh. A.: Russ. J. Inorg. Chem. *25,* 1656 (1980)
38. Cotton, F. A., Gage, L. D., Rice, C. E.: Inorg. Chem. *18,* 1138 (1979)
39. Koz'min, P. A., Surazhskaya, M. D., Larina, T. B.: Russ. J. Inorg. Chem. *26,* 57 (1981)
40. Kuz'min, A. I., Shtemenko, A. V., Kotel'nikova, A. S.: ibid. *25,* 1662 (1980)
41. Eastland, G. W., Yang, G., Thompson, T.: Meth. and Find. Exptl. Clin. Pharmacol. *5,* 435 (1983)
42. Martin, D. S., Huang, H.-W., Newman, R. A.: Inorg. Chem. *23,* 699 (1984)
43. Srinivasan, V., Walton, R. A.: ibid. *19,* 1635 (1980), and references cited therein
44. Brown, P. R., Cloke, F. G. N., Green, M. L. H., Tovey, R. C.: J.C.S. Chem. Commun. 1982, 519
45. Ebner, J. R., Walton, R. A.: Inorg. Chem. *14,* 1987 (1975)
46. Glicksman, H. D., Walton, R. A., Inorg. Chim. Acta *19,* 91 (1976)
47. Bursten, B. L., Cotton, F. A., Fanwick, P. E., Stanley, G. G., Walton, R. A.: J. Am. Chem. Soc. *105,* 2606 (1983)
48. Cotton, F. A., Dunbar, K. R., Falvello, L. R., Tomas, M., Walton, R. A.: ibid. *105,* 4950 (1983)
49. Dunbar, K. R., Walton, R. A.: Inorg. Chem. *24*, 5 (1985)
50. Cutler, A. R., Walton, R. A.: Inorg. Chim. Acta, in press
51. Walton, R. A.: ACS Symp. Ser. *No. 155,* 207 (1981)
52. Cole, N. F., Cotton, F. A., Powell, G. L., Smith, T. J.: Inorg. Chem. *22,* 2618 (1983)
53. Puddephatt, R. J.: Chem. Soc. Revs. 1983, 99
54. Barder, T. J., Cotton, F. A., Lewis, D., Schwotzer, W., Tetrick, S. M., Walton, R. A.: J. Am. Chem. Soc. *106,* 2882 (1984)
55. Abbott, E. H., Bose, K. S., Cotton, F. A., Hall, W. T., Sekutowski, J. C.: Inorg. Chem. *17,* 3240 (1978)

56. Barder, T. J., Cotton, F. A., Dunbar, K. R., Powell, G. L., Schwotzer, W., Walton, R. A.: Inorg. Chem., in press
57. Shaik, S., Hoffmann, R., Fisel, R. C., Summerville, R. H.: J. Am. Chem. Soc. *102,* 4555 (1980)
58. Barder, T. J., Cotton, F. A., Powell, G. L., Tetrick, S. M., Walton, R. A.: ibid. *106,* 1323 (1984)
59. Barder, T. J., Tetrick, S. M., Walton, R. A., Cotton, F. A., Powell, G. L.: ibid. *105,* 4090 (1983)
60. Cotton, F. A., Ilsley, W. H., Kaim, W.: ibid. *102,* 1918 (1980)
61. Barder, T. J., Cotton, F. A., Falvello, L. R., Walton, R. A.: Inorg. Chem. *24,* 1258 (1985)
62. Dunbar, K. R., Walton, R. A.: Inorg. Chim. Acta *87,* 185 (1984)
63. Cotton, F. A., Daniels, L. M., Schmulbach, C. D.: ibid. *75,* 163 (1983)
64. Allison, J. D., Wood, T. E., Wild, R. E., Walton, R. A.: Inorg. Chem. *21,* 3540 (1982)
65. Cameron, C. J., Tetrick, S. M., Walton, R. A.: Organometallics *3,* 240 (1984)
66. Cotton, F. A., Daniels, L. M., Dunbar, K. R., Falvello, L. R., Tetrick, S. M., Walton, R. A.: J. Am. Chem. Soc., in press
67. Anderson, L. B., Barder, T. J., Walton, R. A.: Inorg. Chem., in press
68. Sattelberger, A. P., McLaughlin, K. W., Huffman, J. C.: J. Am. Chem. Soc. *103,* 2880 (1981)
69. Santure, D. J., McLaughlin, K. W., Huffman, J. C., Sattelberger, A. P.: Inorg. Chem. *22,* 1877 (1983)
70. Cotton, F. A., Wang, W.: ibid. *21,* 3860 (1982)
71. Cotton, F. A., Wang, W.: ibid. *23,* 1604 (1984)
72. Chisholm, M. H., Chiu, H. T., Huffman, J. C.: Polyhedron *3,* 759 (1984)
73. Bancroft, G. M., Pellach, E., Sattelberger, A. P., McLaughlin, K. W.: J.C.S. Chem. Commun. 1982, 752
74. Sattelberger, A. P.: ACS Symp. Ser. *No. 211,* p. 291 (1983)
75. Comments delivered by D. L. Lichtenberger following the presentation of the paper in Ref. 74; see ACS Symp. Ser. *No. 211,* p. 301 (1983). Note that through an inadvertant omission, the author of the discussion on p. 301 (Dr. Lichtenberger) is not cited
76. Santure, D. J., Huffman, J. C., Sattelberger, A. P.: Inorg. Chem. *23,* 938 (1984)
77. Girolami, G. S., Mainz, V. V., Andersen, R. A.: ibid. *19,* 805 (1980)
78. Cotton, F. A., Lay, D. G.: ibid. *20,* 935 (1981)
79. Girolami, G. S., Andersen, R. A.: ibid. *21,* 1318 (1982)
80. Cotton, F. A., Kalbacher, B. J.: ibid. *16,* 2386 (1977)
81. Sharp, P. R., Schrock, R. R.: J. Am. Chem. Soc. *102,* 1430 (1980)
82. Schrock, R. R., Sturgeoff, L. G., Sharp, P. R.: Inorg. Chem. *22,* 2801 (1983)
83. Cotton, F. A., Felthouse, T. R., Lay, D. C.: J. Am. Chem. Soc. *102,* 1431 (1980)
84. Cotton, F. A., Extine, M. W., Felthouse, T. R., Kolthammer, B. W., Lay, D. G.: ibid. *103,* 4040 (1981)
85. Cotton, F. A., Felthouse, T. R.: Inorg. Chem. *20,* 3880 (1981)
86. Zietlow, T. C., Klendworth, D. D., Nimry, T., Salmon, D. J., Walton, R. A.: ibid. *20,* 947 (1981)
87. Harwood, W. S., Walton, R. A.: unpublished results
88. Cotton, F. A., Falvello, L. R., Mott, G. N., Schrock, R. R., Sturgeoff, L. G.: Inorg. Chem. *22,* 2621 (1983)
89. Cotton, F. A., Mott, G. N., Schrock, R. R., Sturgeoff, L. G.: J. Am. Chem. Soc. *104,* 6781 (1982)
90. McCarley, R. E., Ryan, T. R., Toradi, C. C.: ACS Symp. Ser. *No. 155,* p. 41 (1981)
91. Ryan, T. R., McCarley, R. E.: Inorg. Chem. *21,* 2072 (1982)
92. McCarley, R. E.: Phil. Trans. R. Soc. Lond. *A308,* 141 (1982)
93. Cotton, F. A., Hubbard, J. L., Lichtenberger, D. L., Shim, I.: J. Am. Chem. Soc. *104,* 680 (1982)
94. Cotton, F. A., Falvello, L. R., Han, S., Wang, W.: Inorg. Chem. *22,* 4106 (1983)
95. Chakravarty, A. R., Cotton, F. A., Shamshoum, E. S.: ibid. *23,* 4216 (1984)
96. DeMarco, D., Nimry, T., Walton, R. A.: ibid. *19,* 575 (1980)
97. Anderson, L. B., Cotton, F. A., DeMarco, D., Fang, A., Ilsley, W. H., Kolthammer, B. W. S., Walton, R. A.: J. Am. Chem. Soc. *103,* 5078 (1981)
98. Cotton, F. A., Mott, G. N.: ibid. *104,* 5978 (1982)

99. Cotton, F. A., Felthouse, T. R., Lay, D. G.: Inorg. Chem. *20,* 2219 (1981)
100. Klendworth, D. D., Welters, III, W. W., Walton, R. A.: Organometallics *1,* 336 (1982)
101. Cotton, F. A., Darensbourg, D. J., Kolthammer, B. W. S.: J. Organomet. Chem. *217,* C14 (1981)
102. Yamamoto, T., Imaizumi, K., Jurata, Y.: Inorg. Chim. Acta *85,* 175 (1984)
103. Collman, J. P., Barnes, C. E., Collins, T. J., Brothers, P. J., Gallucci, J., Ibers, J. A.: J. Am. Chem. Soc. *103,* 7030 (1981)
104. Collman, J. P., Barnes, C. E., Swepston, P. N., Ibers, J. A.: ibid. *106,* 3500 (1984)
105. No journal report on this compound has ever appeared, but see MBBMA, p. 197 for information
106. Clegg, W.: Acta Crystallogr. *B36,* 3112 (1980)
107. Kimura, T., Sakurai, T., Shima, M., Togano, T., Mukaida, M., Nomura, T.: Bull. Chem. Soc. Japan *55,* 3927 (1982)
108. Malinski, T., Chang, D., Feldmann, F. N., Bear, J. L., Kadish, K. M.: Inorg. Chem. *22,* 3225 (1983)
109. Chavan, M. Y., Feldmann, F. N., Lin, X. Q., Bear, J. L., Kadish, K. M.: ibid. *23,* 2373 (1984)
110. Chakravarty, A. R., Cotton, F. A., Tocher, D. A.: J. Am. Chem. Soc. *106,* 6409 (1984)
111. Chakravarty, A. R., Cotton, F. A., Tocher, D. A.: Inorg. Chem. *24,* 172 (1985)
112. Tooze, R. P., Motevalli, M., Hursthouse, M. B., Wilkinson, G.: J.C.S. Chem. Commun. 1984, 799
113. Cotton, F. A., Thompson, J. L.: J. Am. Chem. Soc. *102,* 6437 (1980)
114. Moore, D. S., Alves, A. S., Wilkinson, G.: J.C.S. Chem. Commun. 1981, 1164
115. Stephenson, T. A., Tocher, D. A., Walkinshaw, M. D.: J. Organomet. Chem. *232,* C51 (1982)
116. Cotton, F. A., Chakravarty, A. R., Tocher, D. A., Stephenson, T. A.: Inorg. Chim. Acta *87,* 115 (1984)
117. Behling, T., Wilkinson, G., Stephenson, T. A., Tocher, D. A., Walkinshaw, M. D.: J.C.S. Dalton Trans. 1983, 2109
118. Chakravarty, A. R., Cotton, F. A., Tocher, D. A.: J.C.S. Chem. Commun. 1984, 501; Inorg. Chem. *23,* 4697 (1984)
119. Chakravarty, A. R., Cotton, F. A., Tocher, D. A., Tocher, J. H.: Organometallics *4,* 8 (1985)
120. Chakravarty, A. R., Cotton, F. A., Tocher, D. A.: Inorg. Chem. *24,* 1334 (1985)
121. Chakravarty, A. R., Cotton, F. A., Tocher, D. A.: ibid. *23,* 4693 (1984)
122. Tetrick, S. M., Coombe, V. T., Heath, G. A., Stephenson, T. A., Walton, R. A.: ibid. *23,* 4657 (1984)
123. Collman, J. P., Barnes, C. E., Woo, L. K.: Proc. Natl. Acad. Sci. USA *80,* 8784 (1983)
124. For a general, introductory explanation of the principles, see Levy, D. H.: Scientific American 250 (No. 2), 96 (1984)
125. Morse, M. D., Smalley, R. E.: Ber. Bunsenges. Chem. *88,* 228 (1984)
126. Hopkins, J. B., Langridge-Smith, P. R. R., Morse, M. D., Smalley, R. E.: J. Chem. Phys. *78,* 1627 (1983)
127. Goodgame, M. M., Goddard, W. A., III: Phys. Rev. Letters *48,* 135 (1982)
128. Michalopoulos, D. L., Gensic, M. E., Hansen, S. G., Powers, D. E., Smalley, R. E.: J. Phys. Chem. *86,* 3914 (1982)
129. Riley, S. J., Parks, E. K., Pobo, L. G., Wexler, S.: J. Chem. Phys. *79,* 2577 (1983)
130. Pellin, M. J., Gruen, D. M.: ibid. *79,* 5887 (1983)
131. Bondybey, V. E., English, J. H.: Chem. Phys. Letters *94,* 443 (1983)
132. Efremov, Y. M., Samoilova, A. N., Gurvich, L. V.: Opt. Spectrosc. *36,* 381 (1974)
133. Kok, R. A., Hall, M. B.: J. Phys. Chem. *87,* 715 (1983)
134. Walch, S. P., Bauschlicker, C. W., Jr., Roos, B. O., Nelin, C. J.: Chem. Phys. Letters *103,* 175 (1983)
135. Delley, B., Freeman, A. J., Ellis, D. E.: Phys. Rev. Letters *50,* 488 (1983)
136. Langridge-Smith, P. R. R., Morse, M. D., Hansen, G. P., Smalley, R. E., Merer, A. J.: J. Chem. Phys. *80,* 595 (1984)
137. Cotton, F. A., Shim, I.: J. Am. Chem. Soc. *104,* 7025 (1982)
138. Cotton, F. A., Wang, W.: Nouv. J. Chem. *8,* 331 (1984)
139. Atha, P. M., Hillier, I. H., Guest, M. F.: Mol. Phys. *46,* 437 (1982)
140. Atha, P. M., Hillier, I. H., MacDowell, A. A., Guest, M. F.: J. Chem. Phys. *77,* 195 (1982)

141. Correa de Mello, P., Edwards, W. D., Zerner, M. C.: Int. J. Quantum Chem. *23,* 425 (1983)
142. Kok, R. A., Hall, M. B.: J. Am. Chem. Soc. *105,* 676 (1983)
143. Atha, P. M., Campbell, J. C., Garner, C. D., Hillier, I. H., MacDowell, A. A.: J.C.S. Dalton Trans. 1085 (1983)
144. Wiest, R., Benard, M.: Chem. Phys. Letters *98,* 102 (1983)
145. Ketkar, S. N., Fink, M.: J. Am. Chem. Soc. *107,* 338 (1985)
146. Blondel, M.: Ann. Chim. Phys. *8,* 110 (1905)
147. Wöhler, L., Frey, W.: Z. Elektroch. *15,* 132 (1909)
148. Delépine, M.: Compt. Rend. *150,* 104 (1910)
149. Muraveiskaya, G. S., Orlova, V. S., Evstaf'eva, O. N.: Russ. J. Inorg. Chem. *19,* 561 (1974)
150. Muraveiskaya, G. S., Kukina, G. A., Orlova, V. S., Evstaf'eva, O. N., Porai-Koshits, M. A.: Dokl. Akad. Nauk SSSR *226,* 76 (1976)
151. Bancroft, D. P., Cotton, F. A., Falvello, L. R., Han, S., Schwotzer, W.: Inorg. Chim. Acta *87,* 147 (1984)
152. Muraveiskaya, G. S., Abashkin, V. E., Evstaf'eva, O. N., Golovaneva, I. F., Shchelokov, R. N.: Sov. J. Coord. Chem. (Engl. transl.) *6,* 218 (1980)
153. Cotton, F. A., Falvello, L. R., Han, S.: Inorg. Chem. *21,* 1709 (1982)
154. Conder, H. L., Cotton, F. A., Falvello, L. R., Han, S., Walton, R. A.: ibid. *22,* 1887 (1983)
155. Cotton, F. A., Han, S., Conder, H. L., Walton, R. A.: Inorg. Chim. Acta *72,* 191 (1983)
156. Cotton, F. A., Falvello, L. R., Han, S.: Inorg. Chem. *21,* 2889 (1982)
157. Schagen, J. D., Overbeek, A. R., Schenk, H.: ibid. *17,* 1938 (1978)
158. Barton, J. K., Rabinowitz, H. N., Szalda, D. J., Lippard, S. J.: J. Am. Chem. Soc. *96,* 2827 (1977)
159. Hollis, L. S., Lippard, S. J.: ibid. *103,* 6761 (1981)
160. Hollis, L. S., Lippard, S. J.: Inorg. Chem. *21,* 2116 (1982)
161. Hollis, L. S., Roberts, M. M., Lippard, S. J.: ibid. *22,* 3637 (1983)
162. Che, C. M., Schaefer, W. P., Gray, H. B., Dickson, M. K., Stein, P. B., Roundhill, D. M.: J. Am. Chem. Soc. *104,* 4253 (1982)
163. Chisholm, M. H., Folting, K., Huffman, J. C., Rothwell, I. P.: ibid. *104,* 4389 (1982)
164. Chisholm, M. H., Folting, K., Hoffman, D. M., Huffman, J. C., Leonelli, J.: J.C.S. Chem. Commun. 1983, 589
165. Chisholm, M. H., Hoffman, D. M., Huffman, J. C.: J. Am. Chem. Soc. *106,* 6806 (1984)
166. Schrock, R. R., Listemann, M. L., Sturgeoff, L. G.: ibid. *104,* 4291 (1982)
167. Chisholm, M. H., Hoffman, D. M., Huffman, J. C.: Inorg. Chem. *22,* 2903 (1983)
168. Chisholm, M. H., Huffman, J. C., Marchant, N. S.: J. Am. Chem. Soc. *105,* 6162 (1983)
169. Cotton, F. A., Schwotzer, W., Shamshoum, E. S.: Organometallics *3,* 1770 (1984)
170. Cotton, F. A., Schwotzer, W., Shamshoum, E. S.: Organometallics *2,* 1167, 1340 (1983)
171. Chisholm, M. H., Folting, K., Heppert, J. A., Hoffman, D. M., Huffman, J. C.: J. Am. Chem. Soc. *107,* 1234 (1985)
172. Lichtenberger, D. L., Blevins, C. H.: J. Am. Chem. Soc. *106,* 1636 (1984)
173. Cotton, F. A.: Chem. Soc. Revs. *12,* 35 (1983)
174. Cotton, F. A., Powell, G. L.: Inorg. Chem. *22,* 1507 (1983)
175. Campbell, F. L., Cotton, F. A., Powell, G. L.: ibid. *23,* 4222 (1984)
176. Cotton, F. A., Fanwick, P. E., Fitch, J. W., Glicksman, H. D., Walton, R. A.: J. Am. Chem. Soc. *101,* 1752 (1979)
177. Ebner, J. R., Tyler, D. R., Walton, R. A.: Inorg. Chem. *15,* 833 (1976)
178. Best, S. A., Smith, T. J., Walton, R. A.: ibid. *17,* 99 (1978)
179. Agaskar, P. A., Cotton, F. A., Fraser, I. F., Peacock, R. D.: J. Am. Chem. Soc. *106,* 1851 (1984)
180. Moscowitz, A.: Tetrahedron *13,* 48 (1961)
181. Mason, S. F.: Molecular Optical Activity and the Chiral Discriminations, Cambridge University Press 1982
182. Snatzke, G.: Angew. Chem. Int. Ed. Engl. *18,* 363 (1979)
183. Snatzke, G., Wagner, U., Wolff, H. P.: Tetrahedron *37,* 349 (1981)
184. Engle, V. J., Gerger, R., Snatzke, G., Wagner, U.: Chem.-Ztg. *105,* 85 (1981)
185. Omelanezuk, J., Snatzke, G.: Angew. Chem. Int. Ed. Engl. *20,* 786 (1981)
186. Cotton, F. A., Falvello, L. R., Murillo, C. A.: Inorg. Chem. *22,* 382 (1983)
187. Han, S.: Ph.D. Dissertation, Texas A & M Univ. 1984
188. Falvello, L. F.: unpublished work, Texas A & M Univ.

Developments in Transition Metal Cluster Chemistry – The Way to Large Clusters –

Günter Schmid

Institut für Anorganische Chemie, Universität Essen, Universitätsstr. 5–7, 4300 Essen 1, F.R.G.

The structure determining rules of transition metal clusters can be divided into two fields of problems. The ligand shell obeys a pattern, following the principle of minimized repulsion on a sphere.

Quantitative descriptions of the chemical bond in large clusters are still unsatisfactory. For small and medium sized clusters some electron counting rules are sensible, whereas for large clusters extended consideration seems to be necessary. Clusters represent intermediates on the way from complexes to metals and hence open a novel line of research. This development is still in its early stages. The effect on different disciplines in chemistry is obvious. Especially large clusters shall serve as models for metal surfaces and give us an insight into the events in catalysis, something which has not been possible until now.

Structure and Bonding 62

1 Introduction

Cluster compounds are fascinating research objects for many scientists in the fields of complex chemistry, solid-state chemistry, physical chemistry and physics. Syntheses, structures, bond-types, physical and chemical properties make transition metal clusters a unique class of chemical compounds. Although fantastic developments in all areas of cluster research could be observed in the course of the last 10–15 years, it looks as if we could still be at the beginning of an evolution, the end of which is not to be foreseen.

Why is there such a fascination for clusters, especially those of the transition metals? Many reasons could be mentioned. One essential fact for the interest is that they occupy a position between molecular and solid-state chemistry. Transition metal clusters are ideal objects in order to study the metal-metal bond running from the single bond to the collective behaviour in the metallic state.

Transition metal clusters are not only of interest in an academic sense, but also in a practical one. This is due to their importance in both homogeneous and heterogeneous catalysis. Small and large transition metal clusters provide manifold opportunities to study important elementary processes in catalysis. So, clarification of such famous catalytic processes as the Fischer-Tropsch-synthesis and others can be expected by synthesizing and investigating various clusters.

There have been several attempts to define the term "metal cluster"[1–7]. Most of the definitions are very similar and the question, should clusters start with a M_2, M_3 or a M_4 unit are really of secondary importance. In this report metal clusters shall be defined as compounds, consisting of three or more metal atoms each of which is connected to at least two other metal atoms of the group by chemical bonds[7]. By this definition, each atom becomes part of a ring (M_3) or a polyhedron (M_4). It agrees in substance with the concept of other authors and seems to be generally accepted. According to this, compounds with a metal-metal bond between two metal atoms, e.g. $(CO)_5Mn–Mn(CO)_5$, will not be considered as metal clusters. Polynuclear metal complexes whose metal atoms are not connected by direct bonds but by ligands, as in the cubic $[FeS(SR)]_4^{n-}$ [8] do not fall within the term "metal cluster" either.

The poorly defined concept of the metal-metal bond[9] sometimes makes it difficult to distinguish between "real" and "false" metal-metal bonds. Many metal-metal bonds are only supposed to exist because of intermetallic distances corresponding to those in the metallic state or because magnetic properties or the noble gas rule "demand" an interaction between neighbouring metal atoms.

The question of the existence of a metal-metal bond mostly arises when metal atoms are connected by bridging ligands. An example of such a "demanded" but experimentally not detected metal-metal bond is found in the case of the well known iron complex $[C_5H_5(CO)_2Fe]_2$[10].

In cluster chemistry a development is indicated promising much more than just a piling up of new compounds: the way leads to an unexplored area in the field of chemical compounds.

We are accustomed to speak of both a molecular chemistry and of a solid-state chemistry. Molecules are understood to be finite and concrete particles in all states of aggregation, consisting of a defined number of atoms, whereas the so-called solid is characterized by its – in atomic dimensions – infinite extension and it normally does not exist in a size corresponding to its formula unit. Clusters are also to be found in solids and are connected, thus forming infinite states[11, 12]. However, the problems to be discussed here consider the growth of small metal clusters to the collective union of atoms. Metal clusters provide us with exceptionally good prerequisites for studying this transition. The transition metal clusters, especially, seem to be a key to solving many problems.

The facts that initiated this report shall be elucidated by Fig. 1.

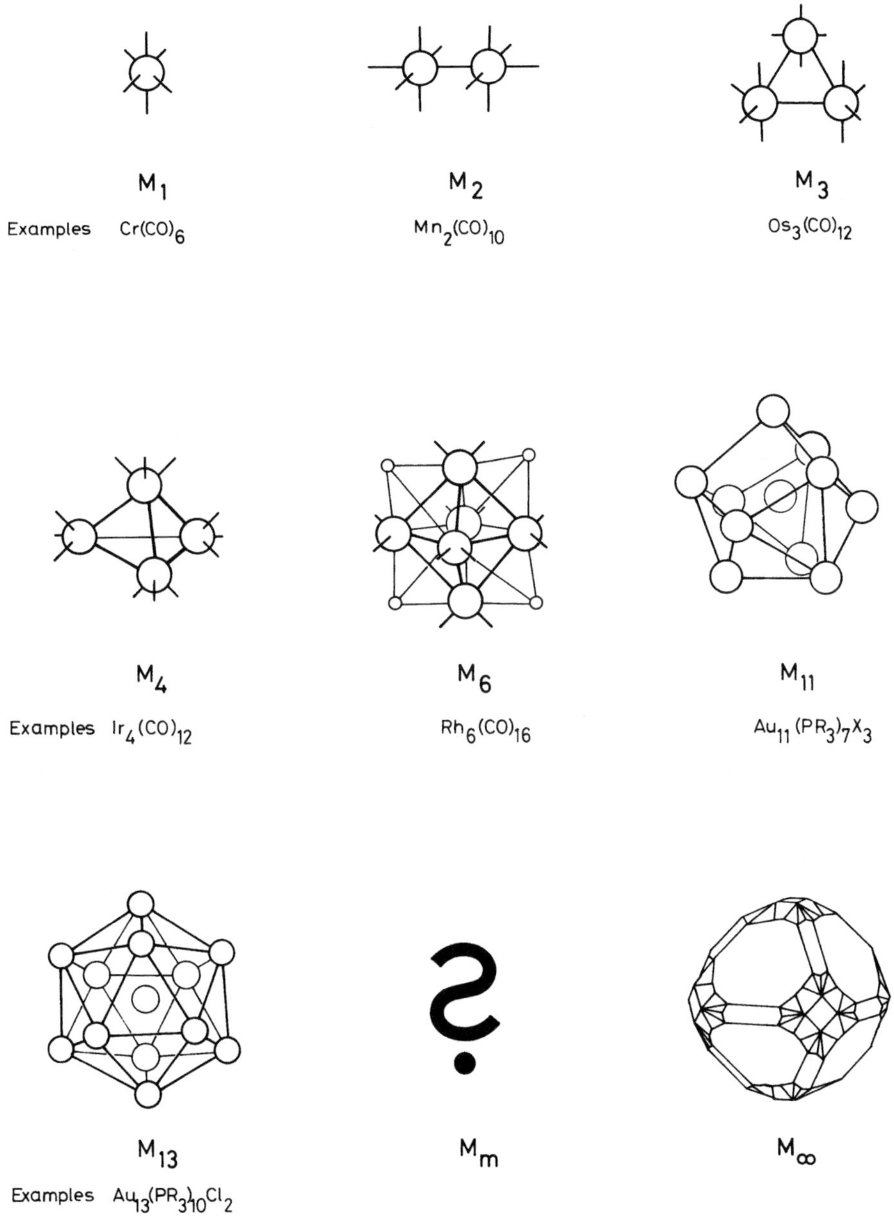

Fig. 1. Examples for mono and binuclear complexes and clusters up to 13 atoms

The examples in Fig. 1 symbolically show the transition from a mononuclear metal complex *via* small clusters with M_3 and M_4 units up to a relatively large cluster with 13 metal atoms. Numerous examples with M_5, M_7, M_8, M_9 or M_{10}, and M_{12} cluster units which are not shown in Fig. 1 have also been described. So, at least from M_1 up to M_{13}, a stepless transition is known. At this point several questions arise:

1. Is it possible to extend this row to a very large m?
2. Which value of m is necessary to bring M_m from a non-metallic to a metallic state?
3. Is there a strategy for the synthesis of clusters with $13 < m < \infty$?
4. Which ligands are necessary to stabilize such large clusters?
5. How are the metal atoms in clusters arranged?
6. Do there exist formations with a special stability and what does this stability depend on?

There are, of course, many more questions, but probably the most essential ones have been mentioned above. Not all of these questions can be answered. For example, there is no answer to the first question. Question 2 is extremely interesting. Essentially it means: how many metal atoms form a metal? We shall refer to this problem later, but it is not surprising that our present knowledge cannot give us a satisfactory answer. The questions 3–6 shall be discussed in the main part of this article.

2 The Ligand Shell

The number and the arrangement of ligands in mononuclear transition metal complexes are normally determined by the electronic and steric conditions, and can be predicted by simple rules. For example, the 18 electron rule and the valence-shell-electron-repulsion (VSEPR) model, respectively, allow us to describe bonding and structure in most of the mononuclear complexes. This may also be possible for small and medium-sized clusters, in so far as they possess covalent 2-center-bonds between the metal atoms and between metal atoms and ligands. Theories of bonding in cluster chemistry are still very unsatisfactory. Mingos and Wade have developed various empirical rules to describe bonding and structure in polyhedral compounds (polyhedral-skeletal-electron-pair-theory = PSEPT)[13–16]. But these rules, which work quite well in borane and carborane chemistry, have a limited value for larger clusters because only s and p orbitals are considered for the cluster bonding. It is assumed that for the later transition metals there is a considerable contraction of the d-orbitals and, as a consequence, little overlap between these orbitals.

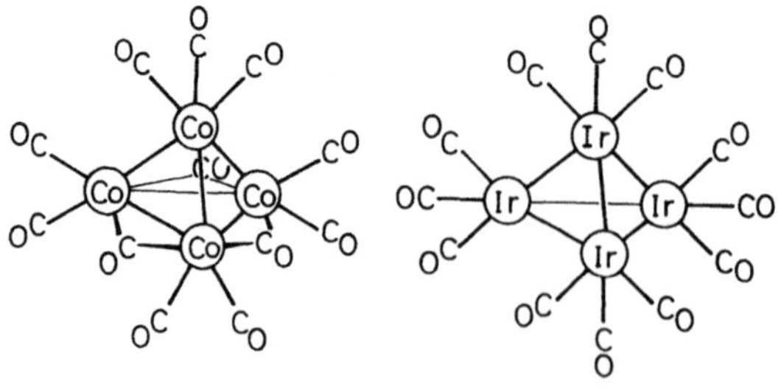

Fig. 2. The molecular structures of $Co_4(CO)_{12}$ and $Ir_4(CO)_{12}$ showing the different ligand shapes

At first, we shall discuss the relationship between the number and the arrangement of ligands in some simple examples. $Co_4(CO)_{12}$ and $Ir_4(CO)_{12}$ are electron precise, tetrahedral clusters with the same empirical formula $M_4(CO)_{12}$, consisting of homologous metals. However, they differ characteristically in structure: in $Co_4(CO)_{12}$ 3 carbonyl bridges connect the 3 basal Co atoms[17], in $Ir_4(CO)_{12}$ only terminal CO groups are observed[18].

The formal substitution of a $Co(CO)_3$ group in $Co_4(CO)_{12}$ by isolobal fragments such as CR[19], SiR[20], GeR[21] or the elements P[22], As[23], S[24, 25] and Se[26] leads to tetrahedral heteronuclear clusters, all having only terminal CO groups co-ordinated to the cobalt atoms. If the heterofragment consists of groups like $(CO)_3Fe$[27], $C_5H_5(CO)_2M$ (M = Mo, W)[28] or $(CO)_3Rh$[29, 30], the CO bridges, known from $Co_4(CO)_{12}$, are once again present. What are the reasons for this?

The structure types (a) and (b) in Fig. 3 differ in the number n of ligands R and in the heteroelement E. The key for understanding the structural differences must be found either in E or in R_n, or in both. A first study shows that CO bridges appear, if E is a transition metal. Only terminal CO ligands are observed if E is a main group element. A further difference between (a) and (b) is due to the number n of the substituents R. If E is a transition element, the favoured co-ordination number of this metal is 6 and therefore n is 3, however it is 1 (for C, Si, Ge) or 0 (for P, As, S, Se) if E is a main group element. For the same number of cluster atoms[4] we have 12 ligands in one case and 9 or 10 ligands in the other case. These differences and their possible structure-decisive roles have been pointed out earlier[7]. Meanwhile, there exist many contributions to this subject. As it seems, both the number of ligands and the size of the cluster nucleus determine the ligand arrangement. Electronic reasons or orbital controlled processes should primarily not be responsible, although this point should not be disregarded. B. F. G. Johnson has made important contributions to this problem[31–33]. Going from $Co_4(CO)_{12}$[17] to $Ir_4(CO)_{12}$[18] or from $Fe_3(CO)_{12}$[34–36] to $Ru_3(CO)_{12}$[37, 38] and $Os_3(CO)_{12}$[37, 38] the carbonyl bridges disappear. Johnson proposed a model according to which the structure of a carbonyl cluster $M_m(CO)_n$ is determined by a polyhedron, built up by the n CO groups, and also by the orientation of the metal cluster within the polyhedron[33]. In $Co_4(CO)_{12}$ and $Fe_3(CO)_{12}$ the 12 CO groups form an icosahedron, in $Ir_4(CO)_{12}$ and $Os_3(CO)_{12}$ they show a cuboctahedral arrangement, although the isosahedron is favoured because of its higher symmetry. The metal clusters Co_4 and Fe_3 fit better into the smaller space of an icosahedron than the bigger Ir_4 or Os_3 nuclei which are better accomodated in a cuboctahedral volume. Thus, only geometrical factors are held responsible for the structures of metal carbonyl clusters. Johnson has extended this model to a series of higher carbonyl clusters. The examples $[Fe_4(CO)_{13}]^{2-}$, $[Co_6(CO)_{14}]^{4-}$, $[Co_4Ni_2(CO)_{14}]^{2-}$, $[Co_6(CO)_{15}]^{2-}$,

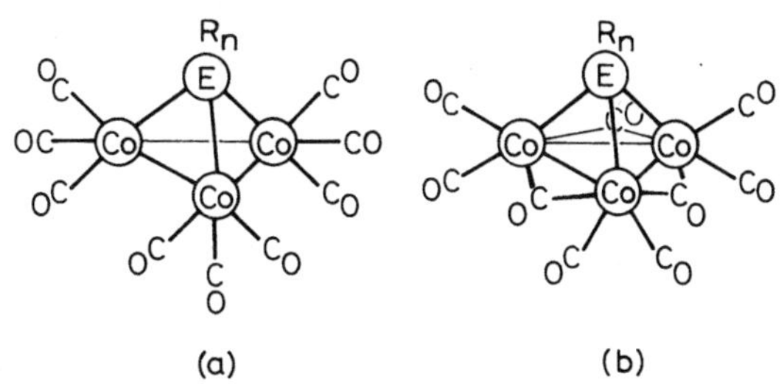

Fig. 3a, b. Structure types of heteronuclear clusters $(CO)_9Co_3ER_n$ · **(a)** with n = 0, 1, **(b)** with n = 3

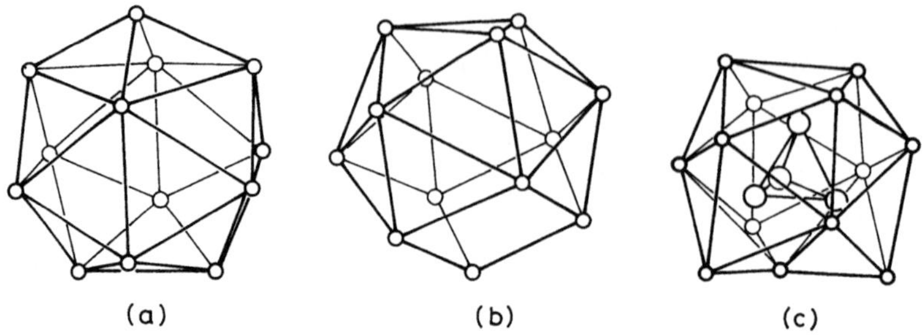

Fig. 4a–c. The best (**a**), the second best (**b**) and the realized (**c**) polyhedron for the 13-vertex polyhedron in $[Fe_4(CO)_{13}]^{2-}$

$[Rh_6C(CO)_{15}]^{2-}$, $[Rh_5Pt(CO)_{15}]^{-}$, $Fe_5C(CO)_{15}$, $Rh_6(CO)_{16}$, $[Fe_6C(CO)_{16}]^{2-}$, $Os_5(CO)_{16}$, $[Mo_2Ni_3(CO)_{16}]^{2-}$, $[W_2Ni_3(CO)_{16}]^{2-}$, and $[Re_4(CO)_{16}]^{2-}$ have been used[33] to compare carbonyl polyhedra, observed by X-ray investigations, with polyhedra which were found by calculation of repulsion potentials between points on the surface of a sphere. The model postulates that the carbonyl groups are arranged in such a way that repulsion is minimized. The cluster nucleus M_m is located symmetrically inside the ligand sphere. Such an approach, the quantitative treatment of which cannot be discussed here in detail, in the last analysis comes to an extended VSEPR model. In this case, the calculation is based on the repulsion potentials of points on a sphere. If we consider the 4, 6 or m metal atoms in a cluster nucleus as one big atom, than the 12, 13 or n CO groups pack as the rules of Gillespie-Nyholm (VSEPR) predict. As mentioned above, 12 ligands are best arranged icosahedrally. If its inside is too small for the cluster M_m, the second best polyhedron, the cuboctahedron, is selected. The best polyhedron according to Johnson's calculation is not found in all cases. Of the 8 most favoured polyhedra for $n = 13$ in $[Fe_4(CO)_{13}]^{2-}$ the second best 13-vertex polyhedron, corresponding to a face-capped icosahedron, is selected and not the energetically more favoured edge-bridged icosahedron.

For $n = 14$ the second best polyhedron according to the calculations, the omnicapped cube, is realized rather than the more favoured bicapped hexagonal antiprism. This second best polyhedron is observed for both $[Co_6(CO)_{14}]^{4-}$ and $[Co_4Ni_2(CO)_{14}]^{2-}$.

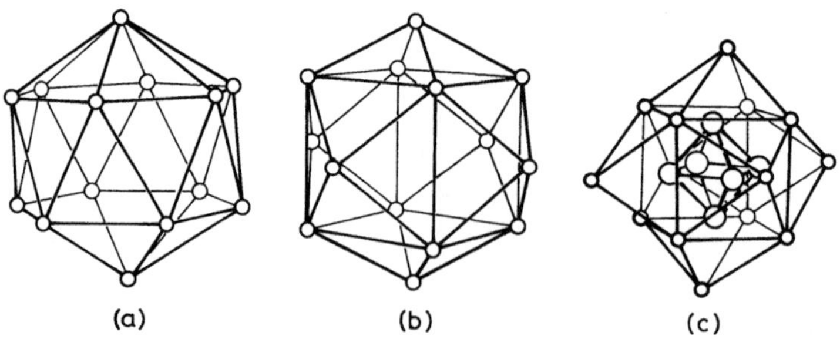

Fig. 5a–c. The best (**a**), the second best (**b**) and the realized (**c**) polyhedron for the 14-vertex polyhedron in $[Co_6(CO)_{14}]^{4-}$ ($[Co_4Ni_2(CO)_{14}]^{2-}$ is isostructural)

For $[Co_6(CO)_{15}]^{2-}$ and $[Rh_6C(CO)_{15}]^{2-}$ the most favoured CO-polyhedron is observed. The second best structure is realized for $[Rh_5Pt(CO)_{15}]^-$ but only the third-best of 6 solutions is selected for the carbonyl clusters $Rh_6(CO)_{16}$, $[Fe_6C(CO)_{16}]^{2-}$ and $Os_5(CO)_{16}$. Concerning this problem, it is to be noted that the differences in energy between the best, the second-best and also the third-best polyhedron, in most cases, are very small. So, the second-best structure for a 16-vertex polyhedron differs by only 0.274%, the third-best solution by only 0.339% from the best polyhedron. With n = 15 the difference between the best and the second-best arrangement is 4.164% and for a 13-vertex structure it is 3.552%. These differences are indeed very small and explain why, in reality, more than one of the energetically quite similar polyhedrons may be built up by the CO groups. Furthermore, it is to be considered that packing effects in the crystal, inaccuracies in the X-ray structure determinations and also a certain influence of the metal nucleus on its sphere may cause deviations from the ideal structure. Nevertheless, these results are in remarkable agreement with the model which considers a cluster unit as a big metal atom, enveloped by ligands which are packed in such a manner as to minimize repulsions between each other. Is the individuality of the single metal atom going to be lost here?

Examples which have a less good agreement between model and experiment should not be concealed: the polyhedra around the Mo_2Ni_3 and W_2Ni_3 nuclei in $[Mo_2(W_2)Ni_3(CO)_{16}]^{2-}$ just as in $[Re_4(CO)_{16}]^{2-}$ belong to the non-favoured structures. The reasons for this may be found in the special metal nuclei Mo_2Ni_3 and W_2Ni_3 with electronically differing metals and Re_4 with the unusual planar co-ordination of the ligands. Besides the steric arguments, it is also argued that there are electronic reasons which may be responsible for the existence or non-existence of carbonyl bridges[39]. The different participation of t_2g-orbitals in M–C–π-bondings is assumed to be responsible for the decreasing tendency in a group of transition metals moving from the light to the heavier elements to form CO bridges. In the case of ruthenium or osmium, e.g. the available t_2g-orbitals are stabilized by π-bonds with terminal CO groups in such a manner that they cannot be used for σ-bonds. The 3 d-orbitals of iron are more contracted and are therefore not available for π-bonds with terminal CO groups such as the d-orbitals in the heavier metals. Furthermore, the formation of a larger number of M–C–σ-bonds is more favoured for the lighter elements, as the individual bonds are weaker, than for the heavier homologues in a group.

We will now return to the problem, discussed at the beginning of this chapter, namely, why clusters of the type $(CO)_9Co_3ER_n$ form carbonyl bridges if ER_n is a fragment such as $Fe(CO)_3$ or $Co(CO)_3$, but contain only terminal CO ligands if E is a main group element with none or only one ligand R, as in the so-called methylidine clusters $(CO)_9Co_3CR$.

The explanations mentioned above make it quite clear that, in the first case, the 12 CO groups form an icosahedron whose vertices occupy the positions of the 9 terminal and the 3 bridging CO groups. 9 or 10 ligands in $(CO)_9Co_3E(R)$ cause another polyhedron without the necessity to occupy bridging positions. If this is correct, then there should be no difference if E is a transition metal or a main group element. To prove this experimentally is rather difficult. Clusters of the type $(CO)_9Co_3ER_n$, where E is either a transition metal with n = 1, or where E is main group element with n = 3, must be synthesized. This can hardly succeed since the transition metals which come into consideration prefer the co-ordination numbers 5 or 6, whilst main group elements have those

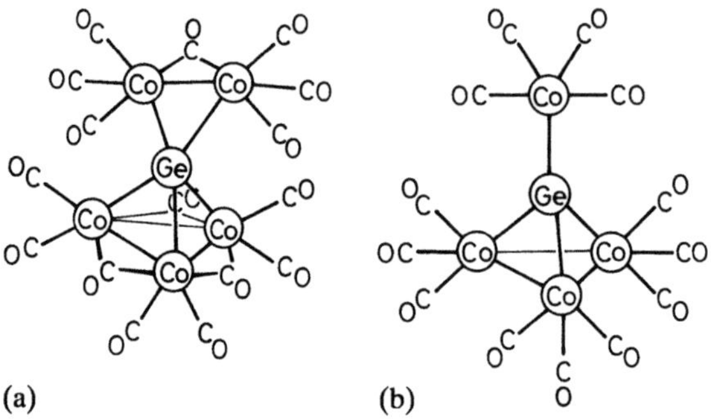

Fig. 6. Molecular structures of $[(CO)_9Co_3GeCo_2(CO)_7]^-$ (**a**) and $(CO)_9Co_3Ge\text{-}Co(CO)_4$ (**b**)

of 3 or 4. Recently B. K. Nicholson and co-workers were able to synthesize the anionic cluster $[GeCo_5(CO)_{16}]^-$ [40]. The X-ray structure shows a tetrahedral $GeCo_3$ unit with five-co-ordinate germanium.

Although the heteroatom Ge in (a) is only five co-ordinate, the Co_3 triangle possesses 3 CO bridges! In contrast to this, the germanium cluster (b) with only four co-ordinate germanium has only terminal CO groups[21]. Not the germanium but the number of ligands around the tetrahedral cluster determines the structure in this case!

In search of an example for a heteronuclear cluster $(CO)_9Co_3ER$ where E is a transition element with only one ligand R, the synthesis of the complex shown in Fig. 7, recently succeeded[41].

Here, the higher co-ordination number of the heteroatom tungsten is avoided by the assumed triple bond to the external tungsten atom. Indeed this complex shows only terminal CO frequencies in the IR spectrum. The compound decomposes rapidly and this is why attempts to investigate it by X-ray methods and to prove the structure proposed in Fig. 7 here failed. The spectroscopic investigations support the proposal.

These examples, as well as Johnson's impressive series of calculated and experimentally found polyhedron shapes for higher carbonyl clusters, allow the following conclusions at the present time: Ligands in the co-ordination sphere of a metal cluster M_m behave in such a manner, that they try to occupy positions on the smallest possible spherical shell, in order to minimize non-bonding repulsions. If the interstice of this shell is too small for a certain cluster nucleus, less favoured arrangements may be accomplished (icosahedron → cuboctahedron when changing from $Co_4(CO)_{12}$ to $Ir_4(CO)_{12}$). The individual character of the single metal atom decreases in its significance with increasing m. The m metal atoms in M_m tend to behave collectively.

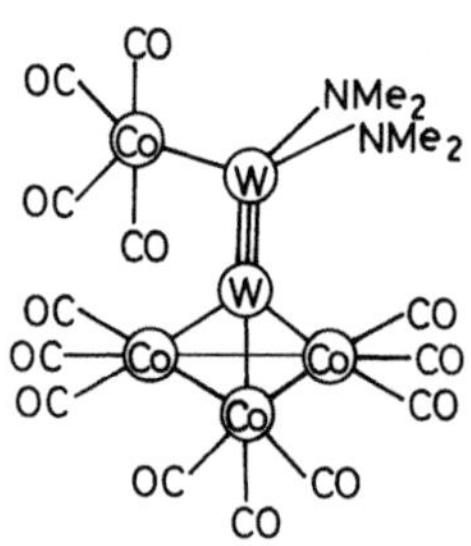

Fig. 7. Proposed structure for $(CO)_9Co_3W[W(NMe_2)_2Co(CO)_4]$

The concept that a metal atom or ion in a complex should be highly saturated co-ordinatively, originates from the classical complex chemistry and is less significant for large clusters. It is more important that the cluster nucleus M_m is enveloped by a ligand shell of appropriate size. New points of view arise if we assume that not each atom on the surface of a cluster M_m must have a certain number of ligands in preferred, orbital controlled directions, but that the novel individuum M_m is to be co-ordinated. A shielding ligand shell may consist either of many small ligands or of a smaller number of large ligands. Consequently, the necessity that each metal atom on the cluster surface co-ordinates one or more ligands possibly becomes inapplicable. We have only to envelope the novel unit M_m. Before we discuss these problems in chapter 6 at some length, we have to deal with the actual cluster nucleus M_m.

3 The Arrangement of the Metals

Tetrahedral or octahedral structures are quite familiar in cluster chemistry and normally do not demand any alternatives. However, the packing of 8, 9 or 10 metal atoms is difficult or impossible to predict without further knowledge. Figure 8 shows some typical examples for packings in metal clusters M_m from $m = 4$ up to 13.

The clusters in Fig. 8 are of course arbitrary, but typical, although structures deviating from these are known. Some of these shall be discussed later.

Both the tetrahedron (M_4), the trigonal bipyramid (M_5) and also the octahedron (M_6) are elements of higher aggregates. Therefore it has been assumed for a long time that metal clusters represent fragments of metallic structures. This may be right on the whole because in clusters and in the metal itself a maximum number of metal-metal bonds is achieved. On the other hand, the finite size of clusters causes special conditions because there are more possibilities to pack the metal atoms, due to the lack of a periodicity. Going from M_4 to M_{13} we observe for the first time at M_8 a metal atom in a central position, surrounded by the other metal atoms in a specific manner. In the Au_8, Au_9 and Au_{11} nuclei this metal atom is the centre of a chair-formed six-membered ring. In the Au_8 case this ring is monocapped, it is bicapped in Au_9, and in Au_{11} 4 additional gold atoms are added. It is easily seen that these geometries consist of triangles and tetrahedra. In the Au_{13} cluster the shape is complete: a central gold atom is co-ordinated icosahedrally by 12 other gold atoms. The number 13 does not result by chance. It is the first of a series of "magic numbers" which were realized in full shell icosahedra or cuboctahedra. M_{13} units are also formed by other metals. $[Rh_{13}(CO)_{24}H_{5-n}]^{n-}$ is another example which is also characterized by a central metal atom surrounded by 12 other metal atoms, in this case in a closest-packed cuboctahedral arrangement. The structure of this Rh_{13} cluster corresponds exactly with a hexagonal closest-packed structure, known for many metals.

Icosahedra and cuboctahedra are closely related. There is still no satisfactory answer to the question, why M_{13} clusters are built up icosahedrally in one case and cuboctahedrally in another case. Possibly the type of ligand or packing effects in the crystal lattice are decisive. The structures, shown in Fig. 8 are all known from the solid state.

In the following, structures of clusters M_m with $m > 13$ shall be considered. Although this row, up to $m = 38$, is rather incomplete, it may be assumed that the structure of larger clusters also follows the principle of closest packings.

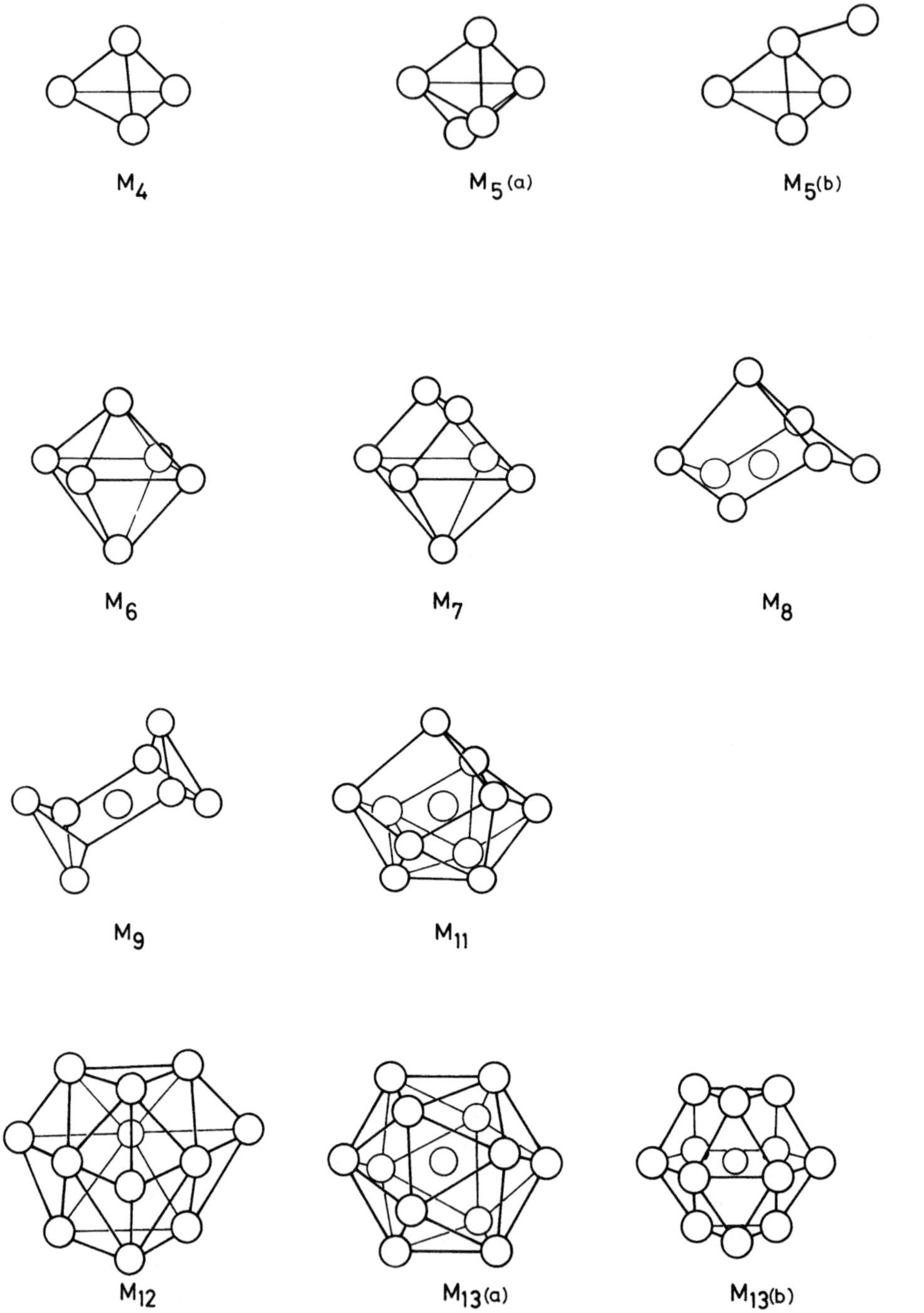

Fig. 8. Selected examples of metal atom packings in metal clusters M_m for m = 4 up to m = 13. Examples: M_4: $M_4(CO)_{12}$ (M = Co[17], Rh[42], Ir[18]); M_5(a): $Os_5(CO)_{16}$[43], $[Ni_5(CO)_{12}]^{2-}$[44]; M_5(b): $[Au_5(diphos)_3]^{2+}$[45]; M_6: $M_6(CO)_{16}$ (M = Co[46], Rh[47]); $Os_6(CO)_{18}$[48]; M_7: $[Rh_7(CO)_{16}]^{3-}$[49]; M_8: $[Au_8(PPh_3)_8]^{2+}$[50]; M_9: $[Au_9(P\,tolyl_3)_8]^{3+}$[52]; M_{11}: $Au_{11}(PR_3)_7X_3$[53–55]; M_{12}: $[Ni_{12}(CO)_{21}H_{4-n}]^{n-}$[56]; M_{13}(a): $[Au_{13}(PR_3)_{10}Cl_2]^{3-}$[57]; M_{13}(b): $[Rh_{13}(CO)_{24}H_{5-n}]^{n-}$[58]

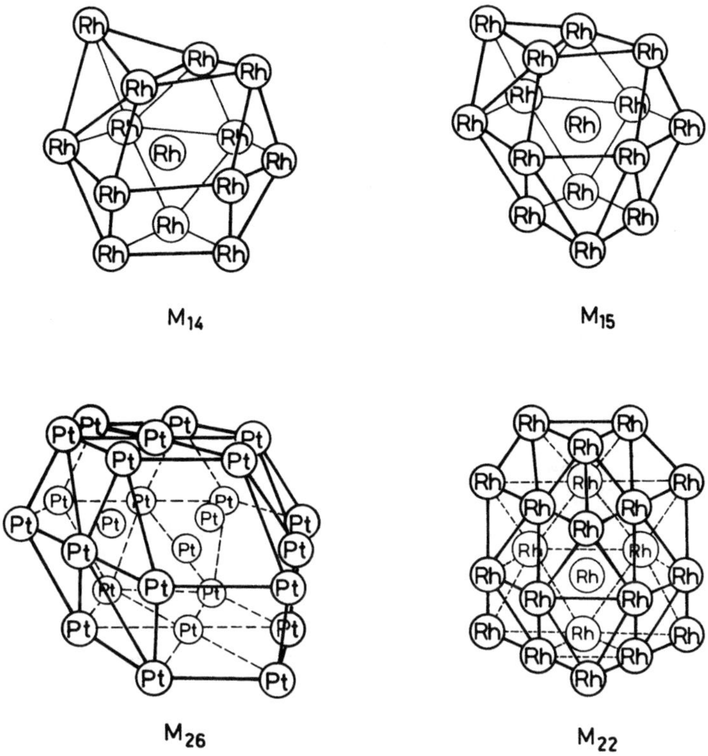

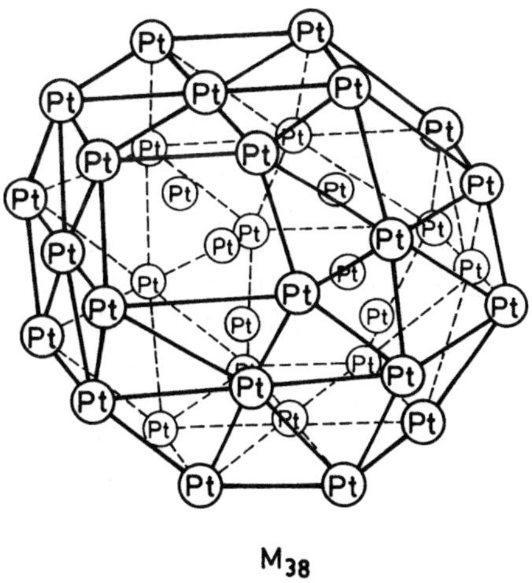

Fig. 9. Examples for metal clusters M_m with $m > 13$: M_{14}: $[Rh_{14}(CO)_{25}]^{4-}$ [59]; M_{15}: $[Rh_{15}(CO)_{27}]^{3-}$ [59]; M_{22}: $[Rh_{22}(CO)_{37}]^{4-}$ [60]; M_{26}: $[Pt_{26}(CO)_{32}]^{2-}$ [61]; M_{38}: $[Pt_{38}(CO)_{44(?)}]^{2-}$ [61]

M_{14} and M_{15} in Fig. 9 differ from M_{13} (hcp) in Fig. 8 by the addition of one or two metal atoms to the cuboctahedron. Consequently, this leads to a decrease of symmetry. The cluster nucleus of $[Rh_{22}(CO)_{37}]^{4-}$ results from the stacking of 4 closest-packed layers with C_{3v}-symmetry. Only one of the 4 layers is completed (7 atoms). The sequence ABAC corresponds to a mixture of hexagonal (hcp) and cubic (ccp) packing.

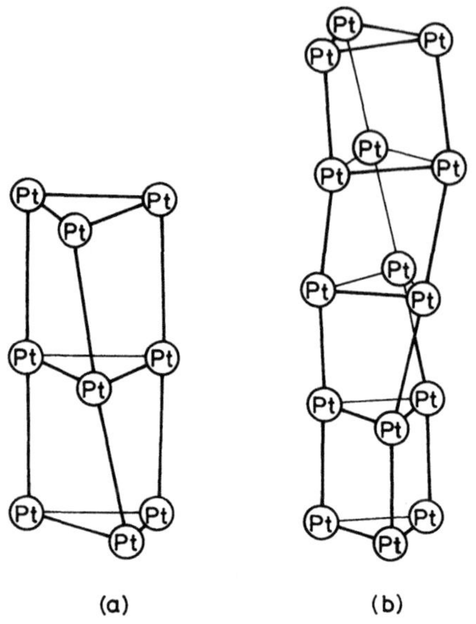

Fig. 10a, b. The metal nuclei of $[Pt_9(CO)_{18}]^{2-}$ (**a**) and $[Pt_{15}(CO)_{30}]^{2-}$ (**b**)

$[Pt_{26}(CO)_{32}]^{2-}$ shows a hcp-structure with the layer sequence ABA, with the 2 layers A consisting of 7 and the middle layer B of 12 platinum atoms. This 12 atom layer is formed when a Pt_7-layer is additionally co-ordinated by 5 further Pt atoms. A preliminary highlight in this development is, no doubt, the synthesis of $[Pt_{38}(CO)_{44}]^{2-}$ [61] having a ccp-structure. The number of the CO ligands could not be determined exactly. The structure of the 38 Pt atoms point out the relationship to compact metals, so that the model which regards clusters as metal fragment becomes more and more convincing.

At this point we have to mention some varying structures. $[Pt_9(CO)_{18}]^{2-}$ [62] and $[Pt_{15}(CO)_{30}]^{2-}$ [62] respectively, consist of Pt_3-triangles which are stacked helically.

The reasons for the oligomerisation of the Pt_3 units are not yet known. The stacking of the triangles also leads to a closest-packed structure; the three-dimensional growth is missing. The structure of $[Pt_{19}(CO)_{22}]^{4-}$ [63] is again not suited to give an icosahedral or cuboctahedral packing by further growth.

The structure consists of three pentagonal bipyramids which are connected *via* vertex atoms. The architecture of the cationic alloy cluster $[(Ph_3P)_{12}Au_{13}Ag_{12}Cl_6]^{m+}$ [63a] (Fig. 11a) is closely related to the Pt_{19} structure. It consists of two Au-centered Au_7Ag_6 icosahedra with a common Au vertex. The atomic arrangement of the 4 last-named examples does not result in a three-dimensional growth to closest-packings, as for the cases mentioned before, but in a stacking of obviously very stable M_3 and M_5 units in one direction. The relatively small number of examples does not allow for any conclusions, whether or not they are exceptions, which might possibly be dependent on the synthetic conditions.

We will now return to clusters with "normal" closest-packed, three-dimensional structures, which may subsequently grow to small crystallites.

Microcrystallites represent highly interesting objects for research in the field of heterogenous catalysis. Big clusters can serve as model compounds for metal crystallites. Questions arising in crystal research may be answered by studying the growth and prop-

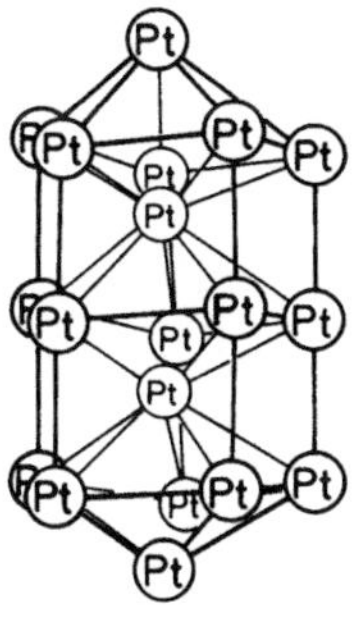

Fig. 11. The metal nucleus of $[Pt_{19}(CO)_{22}]^{4-}$

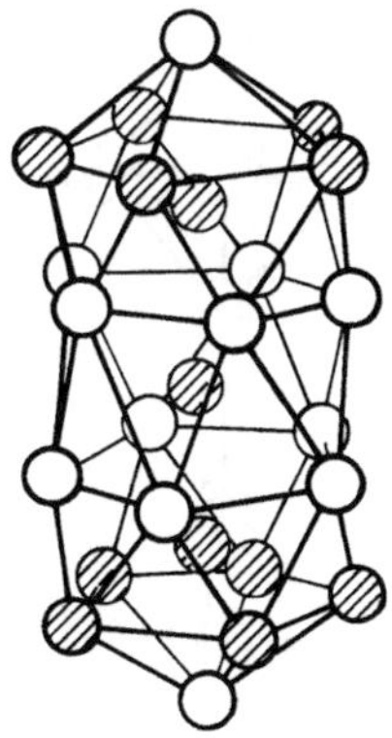

Fig. 11a. The metal nucleus of $[(Ph_3P)_{12}Au_{13}Ag_{12}Cl_6]^{m+}$

erties of such clusters. The connection between ligands in clusters and chemisorbed molecules on a metal surface are obvious. So, for example, the description of interactions between chemisorbed CO and metal atoms on a surface is the same as for the bond between metal atoms and CO in metal carbonyls[64].

During the last few years there were numerous attempts to explain the mechanism of cluster and crystal growth. However, a rigorous differentiation is necessary at this point. The syntheses of the larger clusters described in this chapter occur step by step. This can take place by simple chemical reactions or by less predictable thermal rearrangements and decomposition reactions[61, 65, 66].

$$[Rh_6(CO)_{15}]^{2-} + [Rh(CO)_4]^- \xrightarrow[\text{1 atm CO}]{25\,°C,\ CH_3OH} [Rh_7(CO)_{16}]^{3-} + 3\,CO$$

$$[Rh_{13}(CO)_{24}H]^{4-} + [Rh(CO)_2(CH_3CN)_2]^+ \xrightarrow[CH_3CN]{25\,°C} [Rh_{14}(CO)_{25}H]^{3-} + CO$$

$$Rh_4(CO)_{12} \xrightarrow[N_2]{80\,°C,\ OH^-} [Rh_{15}(CO)_{27}]^{3-}$$

Normally these reactions do not enable the undisturbed growth of a cluster. On the contrary, the cluster growth is influenced by reaction mechanisms, experimental conditions, etc. and, above all, it is unlikely that free metal atoms are formed during these chemical reactions. The structure of $[Rh_{12}(CO)_{30}]^{2-}$[67], consisting of linked octahedra would never be formed, if the 12 Rh atoms might interact undisturbed.

Contrary to this "chemical cluster growth" is the formation of naked, ligand free clusters in a more physical sense. Considering the cluster growth in Fig. 8 and 9, it is to be recognized that both five-fold symmetry and hexagonal (hcp) or cubic (ccp) structures are realized. While in smaller clusters the icosahedral and cuboctahedral symmetry are

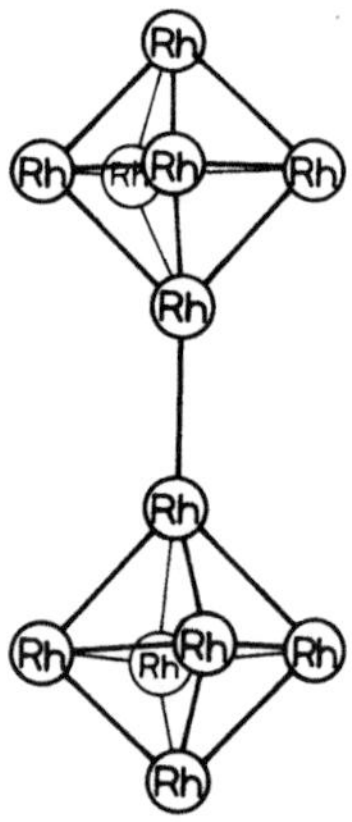

Fig. 12. The metal nucleus of $[Rh_{12}(CO)_{30}]^{2-}$

obviously equivalent, in macroscopic structures the principle of closest-packing is valid. Icosahedral structures do not represent closest-packing arrangements, this is why they are forbidden in macrostructures. Inspite of this they occur occasionally. For example, gold crystals and artificial diamonds with a pentagonal symmetry are known[68, 69]. Many results have been described in the physical literature on crystal growth and on calculated stabilities and instabilities[70–77]. These problems will not be discussed here. It is mentioned that a special stability of full-shell clusters should not be allowed. Furthermore, cluster structures with magic numbers such as 13, 55, 147 etc. should either not exist[78, 79] or should not be observed[80]. Recently this assumption could be disproved experimentally when the existence and size of chemically non-stabilized free clusters was realized for the first time[81]. The mass spectrum of a supersaturated atmosphere of xenon shows definite maxima for 13, 55 and 147 atom xenon clusters. There are also other maxima corresponding to 19, 25, 71 and 87 atom xenon clusters. In contrast, chemically stabilized clusters may be isolated and investigated. The present results show that for smaller and medium-sized clusters not only icosahedral, but also cuboctahedral structures are stable. It is therefore important to clearly distinguish between model clusters and chemically stabilized clusters. Nevertheless, the knowledge in both fields can exert a positive influence on one another. Experiments designed to investigate the nucleation of naked metal clusters arouse more and more interest and numerous uncoordinated metal clusters have been studied by matrix isolation techniques during the last few years[82–85]. Most of the investigated clusters are very small, such as M_2, M_3 or M_4 and are not of direct interest in this report. Some large clusters have also been described, such as Na_{65} which has been indicated by mass spectroscopy with further maxima for Na_8, Na_{19} and Na_{38}[86, 87]. Molybdenum and tungsten clusters with up to 23 and 25 metal atoms[90, 91], respectively, could be synthesized in the gas-phase by supersonic molecular-beam techniques. Trigonal bipyramidal structures have been found ESR-spectroscopically for matrix isolated Cu_5 and Ag_5 clusters[85, 92, 93].

Metal clusters in zeolites can be regarded as intermediate forms between the naked clusters and clusters which are stabilized chemically by a ligand shell. Fe and Co clusters with diameters of 4–11 Å have been prepared from bis(toluol)metal complexes and NaY zeolite[93a].

With respect to the properties of chemically stabilized large clusters, to be discussed in the chapters 6 and 7, our attention should at this point be drawn to some properties of naked clusters. Hoare and Pal proposed that metal clusters of a solid-like nucleus and a liquid-like outer layer with highly mobile surface atoms[94] should exist. This is in contrast to later reports based on solid-like structures[77,95,96]. Later, both cases have been combined by showing that clusters, like macroscopic materials, are well arranged at low temperatures, but that they are liquid-like at higher temperatures through a first order phase transformation[97]. The calculations have been performed for a 55 atom cluster. The transformation from the solid to the liquid state is sharp. However, just before melting, a pre-melting of the cluster surface occurs.

4 Novel Strategies

At the beginning of this report the question arose if, and under which conditions, the row of chemically stabilized, isolable clusters can be continued. Considering the results and conclusions discussed in the chapter above, novel strategies for the synthesis of bigger clusters may now be adapted:

a) Metal atoms in aggregates M_m with $m > 12$ tend to give closest-packed arrangements. We shall not differentiate between the less dense icosahedral or the closest hexagonal or cubic packings for the moment.
b) Closed-shell structures should be due to a special stability.
c) Metal clusters try to get a highly symmetrical ligand envelope.

For the present it seems to be chemically sensible to concentrate our interest on 55 atom clusters as two-layer full-shell structures. In order to achieve controlled synthesis of a M_{55} cluster, it is necessary that the metal atoms are allowed to aggregate unhindered in order to be able to give the closest packing. This has to proceed in the presence of appropriate ligands needed for chemical stabilisation. A local ligand excess prevents cluster growth and favours the formation of smaller clusters and complexes. The most used ligand to date is CO. Its relatively small volume makes it a qualified ligand, as all surface atoms in both small and medium-sized clusters can be co-ordinated either in a terminal or in a bridged manner. The volume of the CO ligand can be determined sufficiently from the shortest O ... O and C ... C distances in metal carbonyls[98].

The distance of parallel CO groups can be estimated to be ~3 Å. As an increasing number of metal atoms is linked with an increased flattening of the cluster surface, in bigger clusters co-ordination problems occur. As a model, we can assume a planar metal surface which is co-ordinated by CO molecules as the closest-packed arrangement. This is reminiscent of the chemisorption of CO to metal surfaces[64]. If we accept an averaged radius of a common transition metal of about 1.2–1.4 Å, this implies that with a parallel CO-arrangement not all of the surface atoms can be co-ordinated.

Similar surface structures should exist in large clusters, beginning with a two-layer 55 atom cuboctahedron. The question arises, whether or not the best coverage of a metal surface can be realized by many small ligands or by fewer bulky ligands, such as phosphines. In contrast to CO, phosphines and related ligands can easily be dosed, in order to prevent formation of smaller clusters.

One strategy adopted to synthesize large full-shell clusters exists in the generation and aggregation of metal atoms in the presence of a few but geometrically appropriate

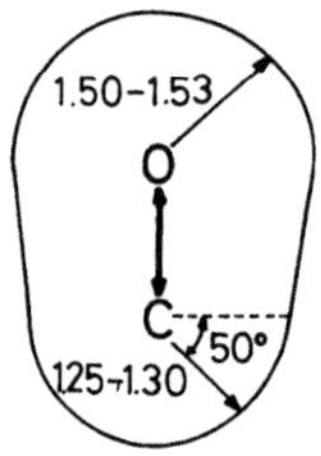

Fig. 13. Contact dimensions of CO (Å)

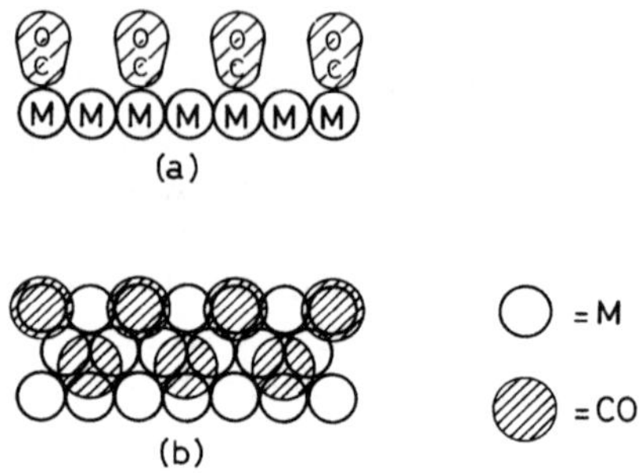

Fig. 14a, b. Model for CO co-ordination on a metal surface (**a**) the space filling of parallel CO groups, (**b**) terminal and bridging CO groups covering a metal surface

ligands. In order to be co-ordinated by these ligands, full-shell structures must have a longer lifetime compared with that of incomplete structures. Concentration and geometry of the ligands must be accomodated to the expected cluster.

The stereochemistry of ligands, especially that of tertiary phosphines is well described by Tolman's cone angle concept[99]. The cone angle describes the angle between the outer substituent atoms, determined by their van der Waals radii, and the metal atom in mononuclear complexes (see Fig. 15).

Vollenbroek[100] and Mingos[101] have transformed Colman's cone angle concept to clusters. The cluster cone angle refers to the cluster centre and is more suited to describe the situation in a cluster. The idealized cone face covers a corresponding plane of the cluster surface. Only one of the metal atoms of this covered plane would interact chemically with a ligand, the other atoms could only be covered mechanically.

The cluster cone angle and the distance l respectively determine the number of ligands and the size of the sphere enveloping the metal cluster.

The smaller δ is, the more ligands are necessary to cover a distinct cluster nucleus. Hence ligand size and cluster size are closely interrelated. For the controlled synthesis of a M_{55} cluster e.g., number and geometry of appropriate ligands can be estimated.

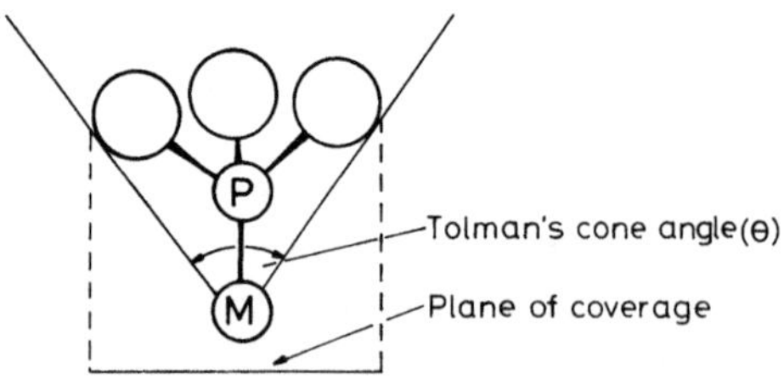

Fig. 15. Tolman's cone angle used in mononuclear complexes with the "plane of coverage"

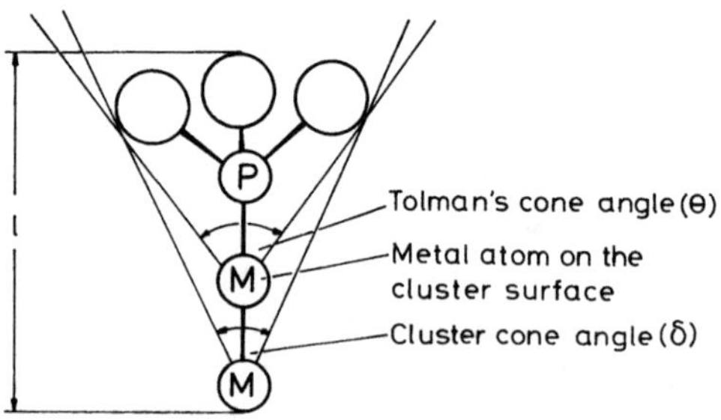

Fig. 16. The cluster cone angle

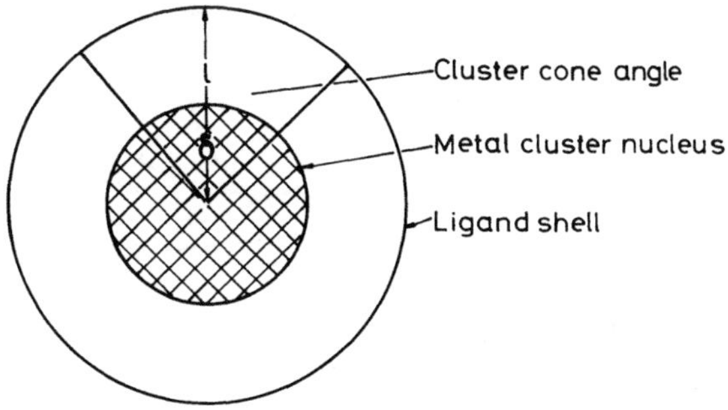

Fig. 17. Schematic construction of the relation between cluster cone angle and ligand shell

Figure 18 gives a schematic representation to evaluate an appropriate ligand for a M_{55} cluster with ccp structure. The proportions in a M_{55} cluster suggest the 12 cuboctahedral (or icosahedral) vertices to be the co-ordination centres.

The correlation between the so-called phosphine radius r_P and the metal atom radius results in the following relationships[102)]

r_P	$= l \cdot \tan 30°$	r_{cub}	= radius of the cuboctahedron,
l	$= r_{cub} + d$	d	= height of the phosphine,
r_{cub}	$= 5 \times r_M$	r_M	= metal atomic radius

As the number of available phosphines or similar ligands is limited, r_P and d have given ratios. One has to try to select d and r_P for a distinct cluster, so that an envelope of touching phosphines results. These conditions obviously cannot exactly be achieved in reality. On the other hand, some defects in the geometrical arrangements should be equalized by the ligands, for these do not consist of rigid balls, but of alkyl or aryl groups which are able to fit into each other. So far, "calculations" such as those mentioned above may give us only approximate values for the ligands which come into consideration.

Figure 19 shows a cuboctahedron which is enveloped by "phosphine circles" in an ideal manner. In each case, 4 touching circles form a gap and the metal atoms below

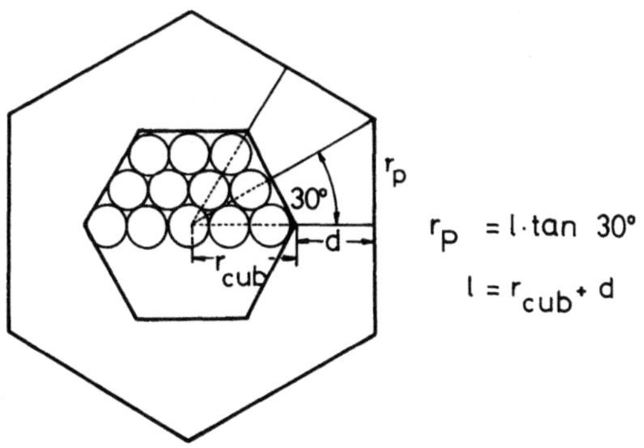

Fig. 18. Model for the "calculation" of an appropriate cluster ligand

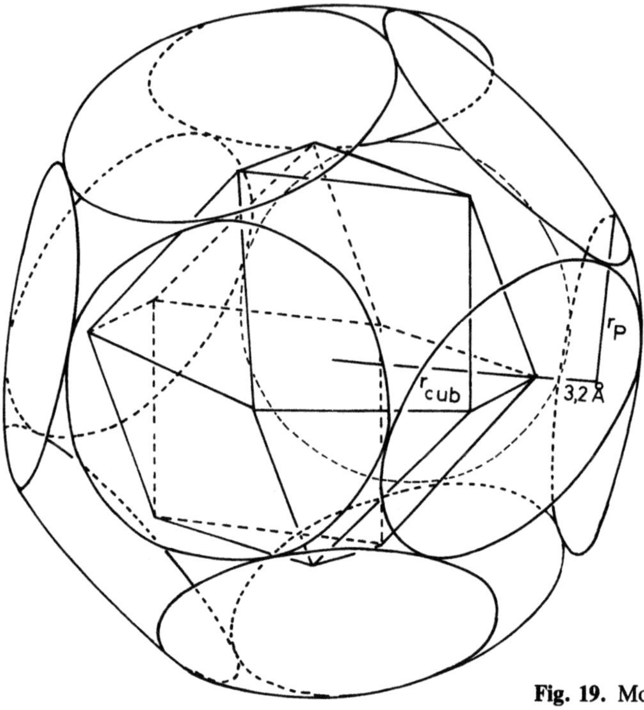

Fig. 19. Model of a molecule $M_{55}(PR_3)_{12}$

remain unshielded. It will depend on the reactivity of the metal as to whether or not these gaps are possible.

How can metal atoms be generated from this point of view? There are two possibilities: from metal salts in solution or by metal vaporization. The co-condensation of gold atoms with R_3PAuX and PR_3 in ethanol yields $[Au_9(PR_3)_8]^{3+}$ and $Au_{11}(PR_3)_7X_3$ (R = para-substituted phenyl group)[100]. $Au_{11}(PPh_3)_7(SCN)_3$ can also be isolated by the co-condensation of gold atoms with PPh_3 and KSCN in ethanol. If gold atoms are evaporated into an ethanol solution of PPh_3 and NH_4PF_6 $[Au_6(PPh_3)_8]^{3+}$, $[Au_8(PPh_3)_8]^{2+}$ and $[Au(PPh_3)_2]^+$ are obtained as a mixture[100]. Until now, larger clusters could not be detected. Certainly the stoichiometries used in these reactions correspond to the formulas of the isolated clusters, e.g. Au : R_3PAuX : PR_3 = 8 : 3 : 4 yields $Au_{11}(PR_3)_7X_3$ and Au : $Au(PR_3)_2X$: PR_3 = 6 : 3 : 2 gives $[Au_9(PR_3)_8]^{3+}$. As mentioned above, a local excess of ligands would prevent the formation of larger clusters. Generating metal atoms in solution is possible by different methods of metal salt reductions which normally lead to black, insoluble metallic particles. Are they formed via closest shell clusters and can they be trapped by appropriate ligands? This question shall be discussed in the next chapter.

5 Novel Large Clusters

If gaseous diborane is passed into a benzene solution of Ph_3PAuCl at 50–60 °C, a compound with the analytical formula $Au_{9.2}(PPh_3)_2Cl$ can be isolated after about 1 h. Molecular weight measurements show the sixfold formula weight leading to the formula $Au_{55}(PPh_3)_{12}Cl_6$. The black solid is soluble in organic solvents such as methylenechloride, chloroform, pyridine, nitrobenzene or tetrahydrofuran giving deep purple coloured solutions. In the course of several days, these solutions decompose to give gold mirrors. The diborane used for the reaction has two functions: it reduces gold(I) in the initial Ph_3PAuCl and it binds excessive PPh_3 to give $H_3B–PPh_3$. This reaction is of decisive importance in this cluster synthesis. In the presence of the whole amount of the starting phosphine, the reduction of Ph_3PAuCl would not lead to a Au_{55} cluster. The reaction of R_3PAuX compounds with $NaBH_4$ proves this. $NaBH_4$ has about the same reduction power as B_2H_6, but only leads to compounds such as $Au_{11}L_7X_3$ and $[Au_{13}(LL)_6]^{4+}$. These are the largest clusters[55, 103, 104]. B_2H_6 equilibrates with various phosphines:

$$B_2H_6 + 2\,PR_3 \rightleftharpoons 2\,H_3B{-}PR_3$$

The position of this equilibrium determines the concentration of free phosphine. The reaction temperature is therefore very important. Under the reaction conditions used, the concentration of free phosphine is obviously just high enough to stabilize the Au_{55} cluster. Triphenylphosphine fulfills the structural requirements discussed in Chap. 4 better than other phosphines or arsines which have also been tested. Evaluations of the dimensions corresponding to the model in Fig. 18 for the Au_{55} cluster result in a good agreement with the experimental data:

$r_{cub} = 5 \times r_{Au} = 7.2$ Å ($r_{Au} = 1.44$ Å from metallic gold)

$d = 3.2$ Å (from X-ray structure determination of Ph_3PAuBr)

$l = 10.4$ Å

From this the radius of the "phosphine circle" is 6.0 Å. The radius found experimentally is 6.25 Å[102]. The difference of 0.25 Å can be adjusted by a denser packing of the phenyl rings. Further dense packings are possible, as shown by the X-ray structures of $Au_{11}[P(p\text{-}FC_6H_4)_3]_7I_3$[55]. The maximum phosphine circle which results from the dimensions of a ligand using the van der Waals radii may be considerably smaller in reality, according to the size of the substituents. The formula $Au_{55}(PPh_3)_{12}Cl_6$ is in complete agreement with the model, whereby the 12 phosphines occupy the 12 vertices of the 55 atom cuboctahedron. The question why the molecule additionally contains 6 chlorine ligands is easily understood if we consider Fig. 19. The 6 gaps formed by the 12 "phosphine circles" are located exactly above the 6 square faces of the cuboctahedron, consisting of 9 metal atoms. If the 6 chlorine atoms are located in the centres of these 6 square faces, these areas are partially closed.

When $Au_{55}(PPh_3)_{12}Cl_6$ is warmed up, especially in solution, it decomposes rapidly according to the following equation:

$$Au_{55}(PPh_3)_{12}Cl_6 \xrightarrow{\Delta T} 6\,(PPh_3)_2AuCl + 49\,Au$$

Free halogen produces a stoichiometric destruction of the cluster:

$Au_{55}(PPh_3)_{12}Cl_6 + 3\ X_2 \rightarrow 6\ Ph_3PAuCl + 6\ Ph_3PAuX + 43\ Au$
(X = Br, I)

The experimental density of $Au_{55}(PPh_3)_{12}Cl_6$ found [3.5 g/cm^3], is in very good agreement with the calculated density of 3.7 g/cm^3. Many experiments aimed at getting visible crystals of $Au_{55}(PPh_3)_{12}Cl_6$ have failed up until now. The same applies to clusters of Rh, Ru and Pt which will be discussed later. In all cases, microcrystalline or amorphous powders are obtained. The reasons for this may be found in the special properties of these molecules (see Chap. 6). X-ray structure determinations could therefore not be carried out. In the case of $Au_{55}(PPh_3)_{12}Cl_6$, Mössbauer-spectroscopic investigations proved helpful to show the correctness of the proposed model[102)]. The cluster model in Fig. 21 shows four different types of gold atoms:

13 central gold atoms in the inner cluster nucleus (A)
24 uncoordinated gold atoms in the surface (B)
12 surface gold atoms, linked with PPh_3 (C)
6 surface gold atoms, linked with Cl (D).

The Mössbauer spectrum is in complete agreement with this assignment[102)], as the parameters correspond to earlier measurements of gold clusters[104–106)]. The isomer shift of line A (− 1.15 mm/s, rel. to ^{197}Pt) is very close to that of metallic gold (− 1.19 mm/s) and is to be assigned to the inner 13 gold atoms. The intensity of the line also verifies this assignment, because the Mössbauer component of these atoms is 2.9 fold higher than for the peripherical gold atoms. Central gold atoms in other clusters show a factor of 2.3. Line B (1.13 mm/s) must be assigned to the uncoordinated surface gold atoms. There are understandably no clusters available for comparison. The lines C (0.88 mm/s) and D (− 0.09 mm/s) result from gold atoms co-ordinated to phosphine and chlorine ligands, respectively[104)].

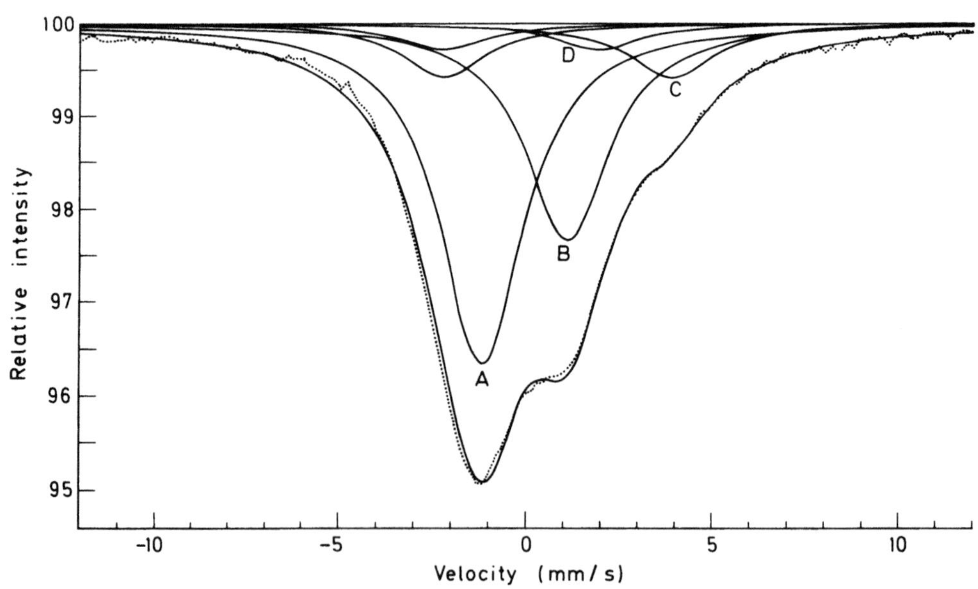

Fig. 20. Mössbauer spectrum of $Au_{55}(PPh_3)_{12}Cl_6$

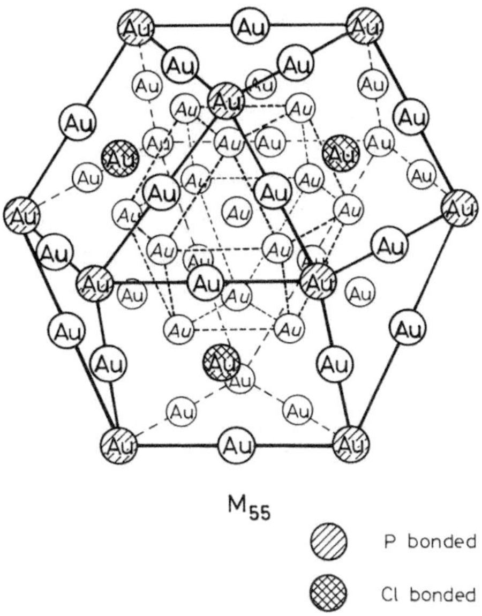

Fig. 21. The Au_{55} cuboctahedron

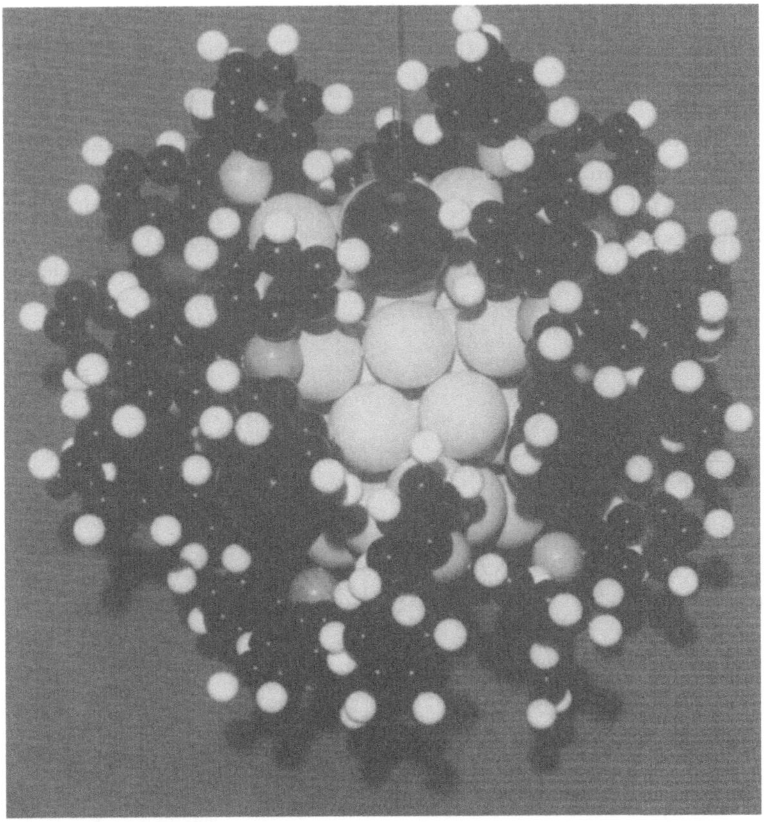

Fig. 22. Space filling atom model of $Au_{55}(PPh_3)_{12}Cl_6$

The unimolecular character of the cluster is apparent from sedimentation measurements[102]. During these measurements, which were carried out to determine the molecular weight in solution, an "optical emptying" of the cell could be observed by a sharp black/white frontier moving to the bottom of the cell. An unhomologeous molecular size would cause an uneven emptying of the cell[107].

The spectroscopic investigations of this and the following clusters shall be discussed in Chap. 6. Figure 22 shows a space filling atom model of $Au_{55}(PPh_3)_{12}Cl_6$.

The enveloping character of the ligand shell is easily seen. If the model for the Au_{55} cluster is correct, it should also be valid for other metals. Indeed, the preparation of the M_{55} clusters of rhodium, ruthenium and platinum succeeded[108, 109]. If PPh_3 were to be selected as a ligand for a Rh_{55} cluster, it would be too bulky with respect to the model in Fig. 18 because Rh is smaller (r_{Rh} = 1.35 Å) than gold (1.44 Å). Under premise that phenyl groups may be packed closer than calculated for an idealized phosphine sphere, PPh_3 might still be suited to Rh.

The reduction of $(Ph_3P)_3RhCl$ with B_2H_6 under the same condition as in the gold case yields $Rh_{55}(PPh_3)_{12}Cl_6$ as a black, extremely air-sensitive powder[108]. The molecular weight determination by sedimentation measurements resulted in the excellent value of 8850 (calculated 9020). As the space filling atom model shows, the phenyl groups on the surface of the Rh_{55} cluster are indeed packed to an almost perfect envelope for the surface atoms.

If a phosphine is used which is in full agreement with the model, namely P(tert-Bu)$_3$, a Rh_{55} cluster is also formed, but it contains more chlorine atoms: Rh_{55}[P(tert-Bu)$_3$]$_{12}Cl_{20}$[109]. If we again consider the space filling atom model of this cluster, it can be seen, that the tert-butyl groups of the phosphine ligands just touch one another, as indicated by model. The reactive Rh atoms bind as many chlorine atoms as possible: one for each triangular face[8] and two for each square[12]. Ru_{55}[P(tert-Bu)$_3$]$_{12}Cl_{20}$ and Pt_{55}[As-(tert-Bu)$_3$]$_{12}Cl_{20}$[108] may be regarded in the same way and confirm that the higher reactivity of these metals, compared with gold, causes a higher chlorine content. Molecular weight determinations again prove the cluster size. Inspite of the close sphere of ligands

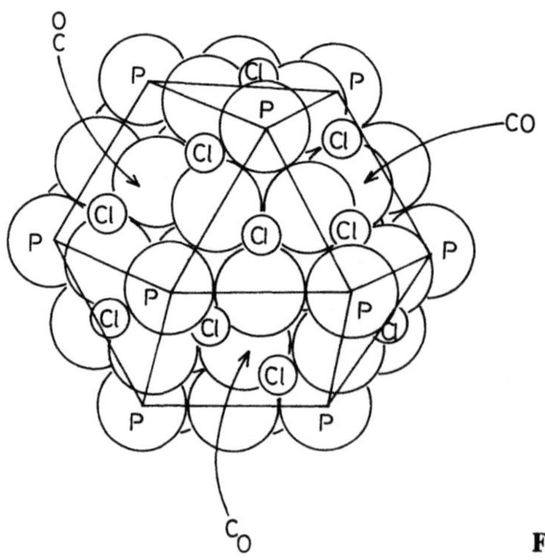

Fig. 23. Addition of CO to Rh_{55}[P(tert-Bu)$_3$]$_{12}Cl_{20}$

on the cluster surface of $M_{55}L_{12}Cl_{20}$ these compounds are still highly reactive. Thus, for example $Rh_{55}[P(tert\text{-}Bu)_3]_{12}Cl_{20}$ can add 6 CO or C_2H_4 molecules in the solid state[108]. Figure 23 shows the possible positions where the 6 CO groups could be co-ordinated.

There are only 6 Rh atoms where this can occur: the Rh atoms of the 6 squares. In agreement with this assumption, only terminal CO groups are observed in the IR spectrum with one peak at 2010 cm^{-1} [109]. These results are not in accordance with icosahedral metal cluster nuclei. Only a cuboctahedral packing makes both the number of chlorine atoms understandable and the exclusively terminal bonded CO groups in $Rh_{55}[P(tert\text{-}Bu)_3]_{12}(CO)_6Cl_{20}$. The question whether hcp or ccp structures are realized, is of secondary importance and cannot be answered at the present time. The ccp structure may be preferable due to the more symmetrical arrangement of the ligands.

6 The Mobility in Clusters

The mobility of ligands in metal cluster molecules and also the non rigidity of the cluster atoms themselves, is a further fascinating property of these compounds. The mobility of ligands on the cluster surface has been shown many times[110]. In most cases, NMR-spectroscopy is the best method to investigate this phenomenon. An NMR-spectrum at normal temperatures describes a time-averaged structure. The temperature dependence of the ligand mobility in carbonyl clusters can be observed by ^{13}C-NMR spectra. Cotton[111, 112] and Johnson[113] were able to prove CO migration in $Rh_4(CO)_{12}$. The low temperature spectrum corresponds to a rigid ligand arrangement because 4 different signals are to be seen in agreement with 4 unequivalent CO groups. The signals are split by coupling of the ^{13}C nuclei with ^{103}Rh nuclei.

If the solution is warmed up, the CO groups begin to move and to change their positions. The time of direct contact is shorter, e.g. the resonance lines begin to broaden. If the contact time of a CO group at a certain Rh atom is comparable with the time required for a nuclear transition (here at 7.5 °C), the signal disappears completely in the noise. Further rise in temperature makes the time of direct contact essentially shorter than the time of the nuclear transition. As a consequence, at 28.6 °C a broad signal is observed, showing the CO ligands and their time averaged environment. If the temperature is increased further, the spectrum shows a fine structure. The fast exchange limit is reached at about 50 °C. It consists of a 1 : 4 : 6 : 4 : 1 quintet with a splitting of 17.1 Hz corresponding to an interaction of a CO group with all 4 Rh atoms. The exchange mechanism is assumed to be a bridge-non-bridge interchange mechnanism. Similar results have been described for $Cp_3Rh_3(CO)_3$[114] or $[Rh_{13}(CO)_{24}H_{5-n}]^{n-}$[115] where hydro-

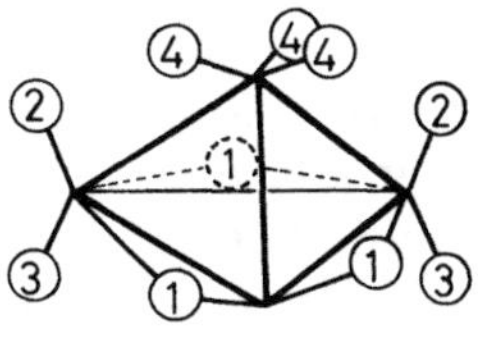

Fig. 24. The structure of $Rh_4(CO)_{12}$ in solid state at low temperature showing 4 different types of CO

gen and carbonyl scrambling is observed. Not all clusters show ligand scrambling, as could be shown for $[Rh_6C(CO)_{15}]^{2-}$[116]. Since all edges are occupied by CO bridges a concerted motion of the CO groups is not possible. Ligand mobility depends on free positions to start scrambling. The mobility in clusters is not limited to the ligands. Little motion of metal atoms can lead to a change in structure. The easy transformation of icosahedral structures into cuboctahedral structures and *vice versa* has been discussed before. Metal motion has been found in $H_2FeRu_3(CO)_{13}$, $H_2FeRuOs_2(CO)_{13}$[117] as well as in $[Rh_9P(CO)_{21}]^{2-}$[118].

Temperature dependent ^{103}Rh-NMR spectra of the Rh cluster show that the Rh skeleton is stereochemically non-rigid. At − 80 °C the spectrum is in agreement with the X-ray structure (3 doublets 1:4:4). At room temperature only one doublet is observed[118].

If the cluster nuclei M_m increase, the mobility of the ligands on the surface should also increase because of the increased flattening. It begins to resemble more and more the behaviour of a metal surface. NMR spectroscopic investigations on M_{55} clusters provide evidence for this. A ^{1}H decoupled ^{31}P NMR spectrum of $Rh_{55}[P(tert\text{-}Bu)_3]_{12}Cl_{20}$ in acetone solution at room temperature shows only one sharp singulet at 68 ppm[109]. ^{31}P–^{103}Rh couplings cannot be observed. $Rh_{55}(PPh_3)_{12}Cl_6$ behaves similarly in CH_2Cl_2 and again only one sharp singulet at 28.5 ppm[108] is observed. These surprising effects show that the phosphine ligands in both clusters are moving at such a pace that only a singulet with completely averaged coupling can be seen. These clusters therefore show "high temperature spectra" at room temperature, a phenomenon which has hitherto not been observed. The exchange has become so fast that $\tau \cdot \Delta\nu(^{31}P) \ll 1$ is valid for the time τ of the longest-lived exchange state. ($\Delta\nu(^{31}P)$ is the largest difference of the ^{31}P larmor frequencies in the various exchange states due to the ^{31}P–^{103}Rh coupling and to the different chemical environment.)

The ^{31}P NMR spectrum of $Rh_{55}(PPh_3)_{12}Cl_6$ in pyridine shows beside the cluster singulet at 28.5 ppm a second small signal at − 4.8 ppm, indicating free triphenylphosphine. It is possible that the empty positions on the cluster surface are occupied by pyridine molecules. The solutions are stable and the cluster can be regained when the pyridine is removed by evaporation. In contrast, the $(tert\text{-}Bu)_3P$ clusters of Rh and Ru are rather unstable in solution. They also dissociate phosphine molecules, but these react with the solvent, probably in a catalytic manner, so that the cluster cannot be regained quantitatively after a certain time. In pyridine, the Ru_{55} cluster even looses the phosphine ligands totally. In the ^{31}P NMR spectrum only free $P(tert\text{-}Bu)_3$ can be registered.

A first experiment, which of course is not yet to be generalized, shows a further surprising property of $Rh_{55}[P(tert\text{-}Bu)_3]_{12}Cl_{20}$. Attempts to registrate a ^{103}Rh spectrum with a special spectrometer, constructed for nuclei with very weak NMR signals[119] such as ^{103}Rh (0.003% of the sensibility of a proton), using the Quadriga technique[120], failed. One of the reasons for this is assumed to be that either the phosphines or the Rh atoms or both species are in a fast motion[109]. This interpretation is in good agreement with the ^{31}P NMR results. $Au_{55}(PPh_3)_{12}Cl_6$ shows only one singulet at 30 ppm in the ^{31}P NMR spectrum as might be expected[102]. From preparative work it is known that $Au_{55}(PPh_3)_{12}Cl_6$ partially dissociates PPh_3 in solution. In contrast to $Rh_{55}(PPh_3)_{12}Cl_6$ a signal for free PPh_3 cannot be observed in the ^{31}P spectrum. Even additional PPh_3, added to a solution of $Au_{55}(PPh_3)_{12}Cl_6$, does not result in a free PPh_3 signal. Furthermore, the cluster signal is shifted to high field with increasing PPh_3 concentrations.

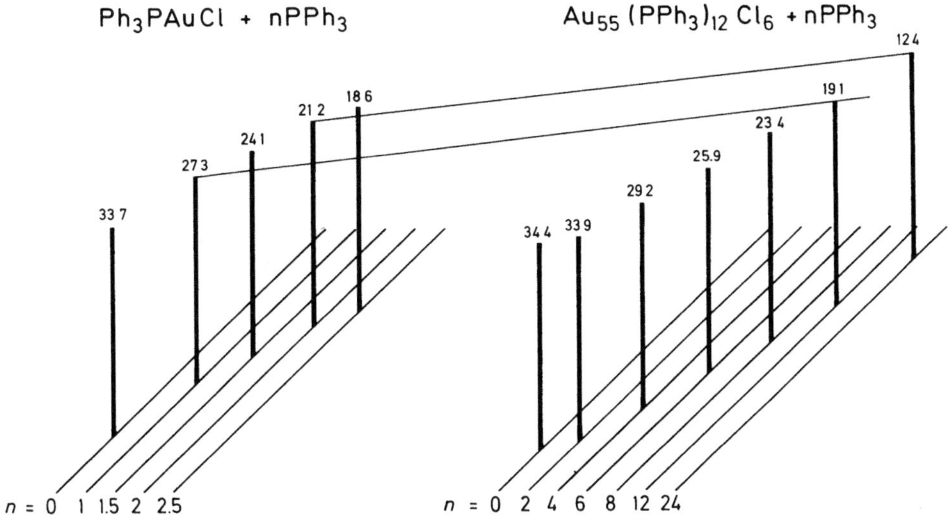

Fig. 25. ^{31}P NMR spectroscopic course of the reaction of Ph_3PAuCl and $Au_{55}(PPh_3)_{12}Cl_6$ with PPh_3 (ppm)

This behaviour is due to a fast exchange between co-ordinated and dissociated phosphines which is averaged by the ^{31}P NMR spectrum. The contact time calculated from these measurements is $\tau = 3.0 \pm 0.5$ µs[121]. A similar qualitative behaviour is found for the system Ph_3PAuCl/PPh_3 but, as can be seen from Fig. 25, there is a large difference in the amount of the effect. If one equivalent of PPh_3 is added to Ph_3PAuCl the ^{31}P signal is shifted to 27.3 ppm; if the number of PPh_3 molecules is doubled in the cluster case the signal moves to 19.1 ppm. A similar relation is found in both cases when two equivalents of PPh_3 are added (21.2 and 12.4 ppm, resp.). This signifies that the ligand exchange in the Au_{55} cluster must be faster than in the simple mononuclear complex. Not all gold clusters show these properties, as can seen at $Au_{11}(PR_3)_7(SCN)_3$[100]. If PPh_3 is added to a solution of this gold cluster in C_6D_6, the signal of free PPh_3 can be observed.

The NMR spectroscopic investigations of small and big clusters can be summarized as follows: The ligands on cluster surfaces are mobile in many of the investigated complexes and change their positions so fast that a definite state of a single ligand cannot be localized by NMR experiments. The large M_{55} clusters allow such fast ligand exchange that even at room temperature couplings between ligands and metal atoms are totally averaged. The mobility of cluster metal atoms may also be assumed, inspite of the fact that only a few examples are known for proving this at the present time. Exchange processes between co-ordinated and dissociated ligands exist with various rates. It should be noted that one physical property of clusters with about 50 metal atoms should be a liquid-like behaviour of the surface atoms[94]. The disorder of the cluster surface in solution may be one reason for the missed tendency to form visible crystals.

7 Bonding in Large Clusters

The classification of clusters as compounds established between complexes and metals leads to the question of how many metal atoms must be present to form a metal. This problem still remains unsolved. The bonding description in clusters of higher nuclearity is still very inadequate. Mingos[122)] and Wade[123)] attempted to combine bonding models of boranes B_pH_{p+q} as developed by Lipscomb[124)], with those for hydrido clusters, using the isolobal relation between BH and $M(CO)_3$ groups. Gillespie[125)] developed rules to describe the relations between the structure of clusters and their number of electrons. They are valid for tetrahedral, octahedral, trigonal bipyramidal, cubic and other medium-sized clusters:

$$N = n\left[18 - 2a\left(\frac{p-1}{p}\right)\right]$$

(n = number of cluster atoms, a = co-ordination number of a single atom, p = co-ordination number of each electron pair.)

For a two-centre bond, p = 2 is valid, for a three-centre bond it is p = 3 etc. This kind of description derives from the well known 18 electron rule. Clusters with central interstitial atoms such as $Ru_6C(CO)_{17}$ or $[Fe_6C(CO)_{16}]^{2-}$ have octahedral structures with an interstitial atom placed at the centre of the octahedron. These compounds are 84 electron clusters with a C^{2+} ion in the centre of 12 cuboctahedrally arranged electron pairs. This bonding consideration also led to the description of gold clusters $[Au_9L_8]^{3+}$ and $[Au_{11}L_{10}]^{3+}$. If it is assumed that in $[Au_9L_8]^{3+}$ each of the peripheral gold atoms makes three electrons available for the cluster bonding, a cuboctahedral structure of 12 electrons is formed. In the centre of this electron cluster there is a Au^{3+} ion, the charge of which corresponds to the total charge of the cluster. The peripheral 8 gold atoms each form four bonds: three bonds to the cluster result and one bond to the ligand is formed. An approximately tetrahedral co-ordination is obtained.

$[Au_{11}L_{10}]^{3+}$ can be described in a similar manner, by the formation of a 10 atom monotruncated cubic cluster which envelopes a Au^{3+} ion. With this model a large number of transition metal clusters can be described, based on two-centre or three-centre bonds. Also clusters such as $[Rh_{13}(CO)_{24}H_3]^{2-}$ and $[Rh_{15}(C)_2(CO)_{14}]^-$ can be characterized by localized two-centre metal-metal bonds using Gillespie's concept. The formation of an electron pair polyhedron with a central atom (metal or carbon) corresponds to the model of a ligand shell around a cluster nucleus. On the basis of the novel compound $[Au_{10}Cl_3(PCy_2Ph)_6]^+$ Mingos developed a structural principle for high nuclearity gold clusters, showing that the Au_{10} framework cannot be regarded as a fragment of an icosahedron[125a)]. Starting with a hypothetical centered chair cluster $[Au_7(PR_3)_6]^+$, two series of gold clusters can develop: those with a total of 12n + 6 electrons and others with 12n + 8 electrons. It is doubted whether these considerations are valid for very large clusters, e.g. of the M_{55} type.

A division into "Wade's" systems, belonging the 2n + 2 rule and "non-Wade's" clusters has been discussed by Struchkov[125b)] with the aim to extend the electron counting procedure to conjuncto-clusters. An additional concept of structural rules for transition metal clusters has recently been presented by Xu[125c)]. Lauher[126, 127)] used extended

Hückel calculations to elucidate the relationship between the number of bonding orbitals in the cluster and the number of metal atoms and their arrangement. Chini[61] summarized the results in the following formula:

number of bonding orbitals $= N \cdot 6 + n$

where the number of bonding orbitals corresponds to half of the cluster valence electrons, N is the number of metal atoms and n is the number of extra bonding high-lying orbitals. n depends on the packing of the metal atoms and varies between 3 and 12. It becomes greater the less dense the packing. Therefore it has its maximum value in the cube. For many of the common structures n lies between 6 and 11. Using Chini's formula for $[Rh_7(CO)_{16}]^{3-}$ the value of n is 7, or it is 3 for the gold cluster $Au_{11}(PR_3)_7I_3$. It follows that n should be negligibly small for infinite structures. As for the heavy transition elements the separation of the p orbitals is large, there are only 6 bonding orbitals (12 electrons) per metal atom, if n is very small. This is the maximum number of orbitals which can be filled in a metal! With that kind of bond description it appears sensible to consider large clusters as parts of a metal lattice.

A simple, but nevertheless very interesting electron counting rule for close-packed high nuclearity clusters has been described by Teo[127a)]. The formula $N = (m_b + rm_s)n$ establishes a connection between the total number of electrons (N), the bulk atoms m_b, and the surface atoms m_s. The rule is based on an extended Hume-Rothery rule and works quite well for all kinds of close-packed clusters up to the Pt_{38} cluster of Chini[61)].

Are these or similar formulas to describe bondings in clusters valid for all kind of clusters or do they have only a limited value? If we use Chini's formula to count electrons in M_{55} clusters, n becomes negative! For example, n is − 12.5 for $Au_{55}(PPh_3)_{12}Cl_6$, − 67.5 for $Rh_{55}(PPh_3)_{12}Cl_6$ and − 60.5 for $Rh_{55}[P(tert\text{-}Bu)_3]_{12}Cl_{20}$. Negative n-values mean that the number of ligands is too small compared with the number of metal atoms. Moreover the number of bonding orbitals depends on the metal one considers, e.g. that for the lighter metals the number of ligands should be higher than for the heavier metals in order to avoid negative n-values. Substitution of many small ligands (e.g. CO) by a smaller number of bulky ligands (e.g. PR_3) should not be possible on this thesis. If Teo's formula is applied to the M_{55} clusters, the rule fails too. The number of observed electrons proves to be much smaller than the calculated number. The reason for this is to be seen in the coordinatively unsaturated cluster surface.

Another approach to describe highly symmetrical metal clusters has been performed by Bacci[127b)], using a "quasi-atomic" picture for the cluster, e.g. in the series $(Nb_6Cl_{12})^{q+}$ $(q = 2, 3, 4)$ the electronic configurations can be expressed as $nf^{14}(n+1)d^{10}(n+2)s^2$ for $q = 2$; $nf^{14}(n+1)d^{10}(n+2)s^1$ for $q = 3$ and $nf^{14}(n+1)d^{10}$ for $q = 4$.

Formulas to count electrons only seem to be available for small and medium-sized clusters. It is apparent that we need other bond descriptions when moving from medium-sized to very large clusters with co-ordinatively unsaturated surfaces.

Molecular orbital calculations on different gold clusters up to 13 gold atoms lead to similar results[128–130)].

A comparison of the MO-levels[129)] of AuL^+ up to $[Au_{13}L_{12}]^{5+}$ shows that the orbitals which are dominantly involved in the metal-metal bonding of this fragment are the hybrid ($6sp_z$) and the higher lying degenerate pair of Au($6p_x$) and Au($6p_y$) orbitals. In agreement with Mingo's calculation[124)], every gold cluster has one lower lying totally symmetric MO which is strongly bonding in the Au-skeleton and shows a stabilization upon

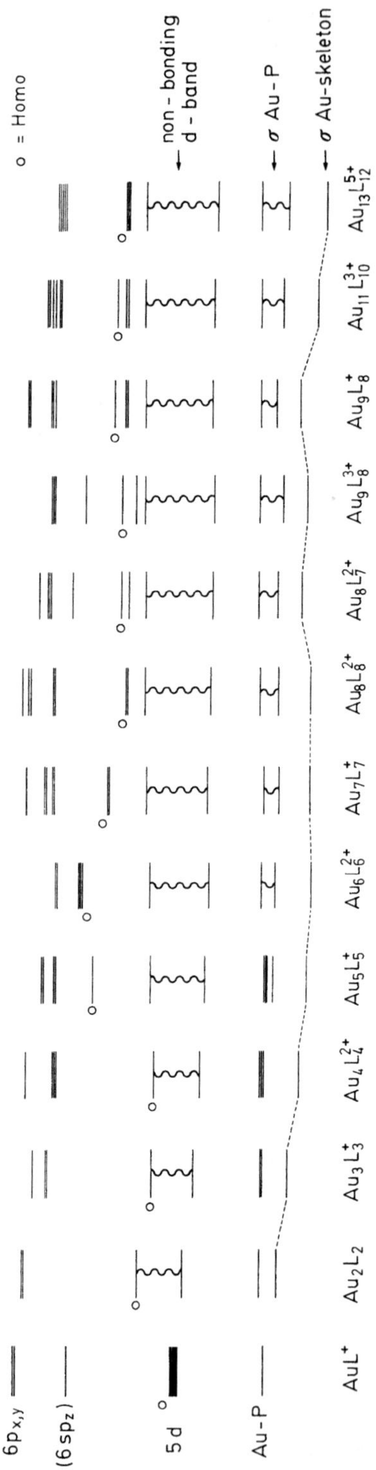

Fig. 26. Relative energy level scheme for a series of gold clusters and the AuL fragment (With the permission of Dr. J. W. A. van der Velden)

increasing cluster size. For clusters with n AuL fragments there exist n MO's representing Au–P bonds. For clusters with n gold atoms 5 n + 1 MO's with essentially non-bonding 5 d-character exist. Empty or partly filled MO's follow, arising from σ and/or π bonding between electrons in 5 d and 6 p gold atomic orbitals. The number of filled MO's varies from 0 to 3 going from 2 to 13 gold atoms. It is possible that the row can be continued up to Au_{55} clusters, but there are no calculations available to date.

Preliminary magnetochemical measurements on $[Au_{13}(diphos)_6](NO_3)_4$ show both in the solid state and in solution a paramagnetism of $\mu_{eff} = 1.1$ BM per formula weight[100]. Single crystals ESR spectra show a broad line without any fine structure. The paramagnetism of $[Au_{13}(diphos)_6](NO_3)_4$ is explained by the presence of an unpaired electron ($Au_{13}^{4+} \equiv 139$ electrons).

Susceptibility measurements on solid $Au_{55}(PPh_3)_{12}Cl_6$[131] in the temperature range of 5.5 K – 294 K show, without diamagnetic correction, up to 25 K very small positive susceptibilities, and above 25 K up to 294 K negative values and indicate diamagnetism of the cluster. This result is supported by negative ESR-experiments. If the susceptibilities are corrected by the diamagnetic values, positive susceptibilities result, to give a magnetic moment $\mu_{eff} = 0.75$ (5.5 K) – ~1.4 BM (294 K). As a net value for the Au_{55} cluster nucleus there remains a paramagnetism corresponding to one unpaired electron. This paramagnetic component is not observable in the very large cluster molecule considering the gross susceptibility. Hence in $Au_{55}(PPh_3)_{12}Cl_6$ we observe diamagnetism considering the whole molecule, whereas for the diamagnetically corrected nucleus one finds one unpaired electron. Metallic gold is known to be diamagnetic. The properties we have observed from M_{55} clusters clearly show that there is a new dimension in cluster chemistry because most of the chemical and physical properties indicate a novel behaviour compared with more simple complexes.

Finally we return to the unusual size of the M_{55} clusters which could be the most important reason for the failure to obtain crystals.

Besides the mobility on the cluster surface in solution the ball-like character of the huge molecules obviously prevents crystallization. If we assume a closest packed structure for the M_{55} cluster sphere in a crystal, then there would exist octahedral lattice vacancies with a diameter of 8–9 Å! There is some hope to find ligands causing intermolecular interactions and thus facilitating crystallization.

By means of high resolution transmission electron microscopy (HRTEM) we recently succeeded in proving the size of the Au_{55} cluster[132].

A drop of a methanolic solution of $Au_{55}(PPh_3)_{12}Cl_6$ was dried at 25 °C on a grid. At a magnification of > 250,000 × instrumental magnification the clusters could be imaged. They are uniformly spread out over the whole graphite film of the grid. The high vacuum of the specimen chamber and the high energy electron beam destroyed most organic material, but the remaining gold clusters exist for a few seconds under these conditions. One complication of this investigation is the growth of larger gold clusters under the influence of the electron beam. The clusters in Fig. 27(a) show a size which is close to the cuboctahedron model, discussed in Chap. 5. The diameter of the magnified cluster is about 7.1 Å (calc. according to Fig. 18, 7.2 Å). As can be seen from Fig. 27(b), the gold clusters grow within 15–20 s to clusters with approximately 20 Å in diameter. The lower part of Fig. 27(b) shows 2 crystals combining to give one crystal. Figure 27(c) shows an almost perfect crystal and it is believed that the perfect crystal growth is due to the "perfect" start from the Au_{55} cluster.

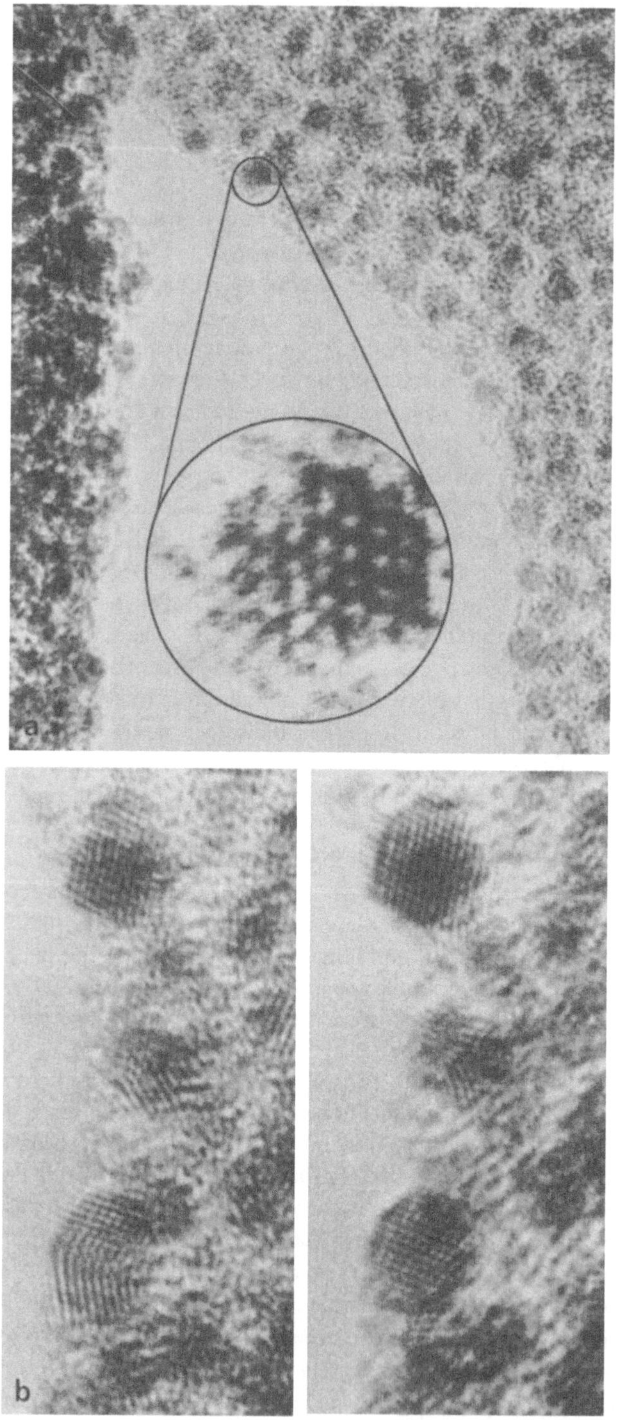

Fig. 27. (**a**) Images of naked gold clusters formed from $Au_{55}(PPh_3)_{12}Cl_6$ by HRTEM (**b**) Gold clusters grown within 15–20 seconds

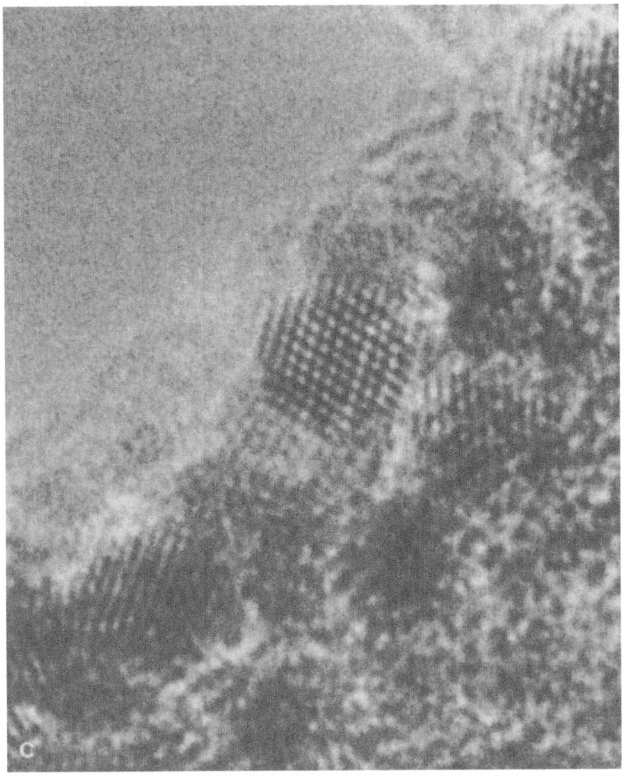

Fig. 27. (c) A perfect gold crystal grown from Au_{55}

These investigations are not yet finished. We hope that by starting with M_{55} clusters crystal growth can be observed step by step for the first time and possibly some of the unsolved problems and questions concerning crystal growth may be solved in the future.

8 Conclusions

The synthesis of transition metal clusters of the type $M_{55}L_{12}X_n$ (X = 6, 20) could be a first step on the way from complexes to metals. There is no doubt that bigger clusters can exist. Those with 147 metal atoms or still higher magic numbers should be characterized by a special stability if an appropriate ligand shell is available as with M_{55} clusters. The higher the number of metal atoms the more probable deviations from the ideal magic number will be. The controlled synthesis of such large, chemically stabilized clusters will become more difficult going from shell to shell. The increasing metal-like character of those clusters complicate isolation and purification of the compounds. Poor solubilities in organic solvent prevent spectroscopic investigations. Finally there will be species which have properties like small metal particles with few organic "impurities". Inspite of these negative aspects, attempts to prepare huge multi-nuclear clusters, for example with 3 or 4 shells should be attempted to test the concept.

The method to prepare M_{55} clusters by diborane reduction of metal complexes is limited, because the reduction potential of B_2H_6 is not strong enough to reduce, e.g., iron, titanium and other less noble metals in their salts. Therefore it is necessary to investigate other reducing agents or other synthetic methods as well. The model to select ligands appears to be useful as does Chini's concept[61)] to use many small ligands for smaller clusters. Chini's other statement, namely that the formation of big clusters succeed better with electron rich metals than with the lighter elements is no doubt correct, because the heavier metals need fewer ligands than the lighter ones.

The super mobile behaviour of the ligands on the M_{55} cluster surfaces in solution is due on the one hand to the flattening of the surface compared with smaller clusters. On the other hand, it is also due to the coordinatively unsaturation of the surface. This aspect seems to be important with respect to elementary catalytic processes.

It is one of the most fascinating points of view, both for academic and for practical reasons, to observe the cluster growth by HRTEM starting with molecular clusters as could be shown with Au_{55}. Never before could cluster growth be studied in such detail.

9 References

1. Cotton, F. A.: Quart. Rev. Chem. Soc. *20,* 389 (1966)
2. Cotton, F. A.: Acc. Chem. Res. *1,* 257 (1968)
3. Kaesz, H. D.: Chem. in Britain *9,* 344 (1973)
4. King, R. B.: Progr. Inorg. Chem. *15,* 287 (1972)
5. Penfold, B. R.: Perspectives in Structural Chemistry *2,* 71 (1968)
6. Vahrenkamp, H.: Structure and Bonding *32,* 1 (1977)
7. Schmid, G.: Angew. Chem. *90,* 417 (1978)
8. Holm, R. H.: Endeavour *34,* 38 (1975)
9. Vahrenkamp, H.: Angew. Chemie *90,* 403 (1978)
10. Mitschler, A., Rees, B., Lehmann, S.: J. Am. Chem. Soc. *100,* 3390 (1978)
11. Simon, A.: Angew. Chem. *93,* 23 (1981)
12. von Schnering, H. G.: ibid. *93,* 44 (1981)
13. Wade, K.: Advan. Inorg. Radiochem. *18,* 1 (1976)
14. Mingos, D. M. P.: Nature, Phys. Sci. *236,* 99 (1972)
15. Mason, R., Mingos, D. M. P.: MTP Int. Rev. Sci. Phys. Chem., Ser. *2,* 11 (1975)
16. Mingos, D. M. P.: J. Chem. Soc. Dalton *1974,* 133
17. Carre, F. H., Cotton, F. A., Frenz, B. A.: Inorg. Chem. *15,* 380 (1976)
18. Churchill, M. R., Hutchinson, J. P.: ibid. *17,* 3528 (1978)
19. Seyferth, D.: Adv. Organomet. Chem. *14,* 97 (1976)
20. Schmid, G., Bätzel, V., Etzrodt, G.: J. Organomet. Chem. *112,* 345 (1976)
21. Boese, R., Schmid, G.: J. Chem. Soc. Chem. Comm. *1979,* 349
22. Vizi-Orosz, A.: J. Organomet. Chem. *111,* 61 (1976)
23. Vizi-Orosz, A., Galamb, V., Palyi, G., Marko, L., Bor, G., Natile, G.: ibid. *107,* 235 (1976)
24. Marko, L., Bor, G., Klumpp, E., Marko, B., Almasy, G.: Chem. Ber. *96,* 955 (1963)
25. Wei, C. M., Dahl, L. F.: Inorg. Chem. *6,* 1229 (1967)
26. Strouse, C. E., Dahl, L. F.: J. Am. Chem. Soc. *93,* 6032 (1971)
27. Chini, P., Colli, L., Peraldo, M.: Gazz. Chim. Ital. *90,* 1005 (1960)
28. Schmid, G., Bartl, K., Boese, R.: Z. Naturforsch. *32B,* 1277 (1977)
29. Martinengo, S., Chini, P., Albano, V. G., Cariati, F., Salvatori, T.: J. Organomet. Chem. *59,* 379 (1973)
30. Johnson, B. F. G., Lewis, J., Matheson, T. W.: J. Chem. Soc. Chem. Comm. *1974,* 441

31. Johnson, B. F. G.: ibid. *1976,* 211
32. Johnson, B. F. G., Benfield, R. E.: J. Chem. Soc. Dalton *1978,* 1554
33. Benfield, R. E., Johnson, B. F. G.: ibid. *1980,* 1743
34. Wei, C. M., Dahl, L. F.: J. Am. Chem. Soc. *88,* 1821 (1966)
35. Wei, C. M., Dahl, L. F.: ibid. *91,* 1351 (1969)
36. Wei, C. M., Wilkes, G. R., Dahl, L. F.: ibid. *89,* 4792 (1967)
37. Mason, R., Rae, A. J. M.: J. Chem. Soc. (A) *1968,* 778
38. Corey, E. R., Dahl, L. F.: Inorg. Chem. *1,* 521 (1962)
39. Evans, D. G.: J. Chem. Soc. Chem. Comm. *1983,* 675
40. Croft, R. A., Duffy, D. N., Nicholson, B. K.: J. Chem. Soc. Dalton *1982,* 1023
41. Bartl, K.: Thesis, Unviversity of Essen 1982
42. Wei, C. M.: Inorg. Chem. *8,* 2384 (1969)
43. Reichert, B. E., Sheldrick, G. M.: Acta Cryst. *B33,* 173 (1977)
44. Longoni, G., Chini, P., Lower, L. D., Dahl, L. F.: J. Am. Chem. Soc. *97,* 5034 (1975)
45. van der Velden, J. W. A., Bour, J. J., Vollenbroek, F. A., Beurskens, P. T., Smits, J. M. M.: J. Chem. Soc. Chem. Comm. *1979,* 1162
46. Albano, V., Chini, P., Scatturin, V.: ibid. *1968,* 163
47. Corey, E. R., Dahl, L. F., Beck, W.: J. Am. Chem. Soc. *85,* 1202 (1963)
48. Mason, R., Thomas, K. M., Mingos, D. M. P.: ibid. *95,* 3802 (1973)
49. Chini, P., Longoni, G., Albano, V. G.: Avd. Organomet. Chem. *14,* 285 (1976)
50. Manassero, M., Naldini, L., Sansoni, M.: J. Chem. Soc. Chem. Comm. *1979,* 385
51. Vollenbroek, F. A., Bosmann, W. P., Bour, J. J., Noordik, J. H., Beurskens, P. T.: ibid. *1979,* 387
52. Bellon, P. L., Cariati, F., Manassero, M., Naldini, L., Sansoni, M.: ibid. *1971,* 1423
53. McPartlin, M., Mason, R., Malatesta, L.: ibid. *1969,* 334
54. Albano, V. G., Bellon, P. L., Manassero, M., Sansoni, M.: ibid. *1970,* 1210
55. Bellon, P. L., Manassero, M., Sansoni, M.: J. Chem. Soc. Dalton *1972,* 1481
56. Broach, R. W., Dahl, L. F., Longoni, G., Chini, P., Schultz, A. J., Williams, J. M.: Adv. Chem. Soc. *167,* 93 (1978)
57. Briant, C. E., Theobald, B. R. C., White, J. W., Bell, L. K., Mingos, D. M. P., Welch, A. J.: J. Chem. Soc. Chem. Comm. *1981,* 201
58. Albano, V. G., Ceriotti, A., Chini, P., Ciani, G., Martinengo, S., Anker, W. M.: ibid. *1975,* 859
59. Martinengo, S., Ciani, G., Sironi, A., Chini, P.: J. Am. Chem. Soc. *100,* 7096 (1978)
60. Martinengo, S., Ciani, G., Sironi, A.: ibid. *102,* 7564 (1980)
61. Chini, P.: J. Organomet. Chem. *200,* 37 (1980)
62. Calabrese, J. C., Dahl, L. F., Chini, P., Longoni, G., Martinengo, S.: J. Am. Chem. Soc. *96,* 2614 (1974)
63. Washecheck, D. M., Wucherer, E. J., Dahl, L. F., Ceriotti, A., Langoni, G., Manassero, M., Sansoni, M., Chini, P.: ibid. *101,* 6110 (1979)
63a. Teo, B. K., Keating, K.: J. Am. Chem. Soc. *106,* 2224 (1984)
64. Ertl, G.: Gazz. Chim. Ital. *109,* 217 (1979)
65. Martinengo, S., Chini, P.: ibid. *102,* 344 (1972)
66. Fumagalli, A., Koetzle, T. F., Tagusagawa, F., Chini, P., Martinengo, S., Heaton, B. T.: J. Am. Chem. Soc. *102,* 1740 (1980)
67. Albano, V. G., Bellon, P. L.: J. Organometal. Chem. *19,* 405 (1969)
68. Allpress, G., Sanders, J. V.: Surface Sci. *7,* 1 (1967)
69. Wantorf, R. H.: The Art and Science of Growing Crystals, J. Gilman Edit., Wiley, New York 1963, p. 176
70. Bagley, B. G.: Nature *208,* 674 (1965)
71. Bagley, B. G.: ibid. *225,* 1040 (1970)
72. Bagley, B. G.: J. Cryst. Growth *6,* 323 (1970)
73. Burton, J. J.: Cat. Rev. Sci. Eng. *9,* 209 (1974)
74. Mackey, A. L.: Acta Cryst. *15,* 916 (1962)
75. Hoare, M. R., Pal, P.: Nature, Phys. Sci. *236,* 35 (1972)
76. Burton, J. J.: Chem. Phys. Lett. *3,* 594 (1970)
77. Burton, J. J.: Nature *229,* 335 (1971)

78. Hoare, M. R., Pal, P.: Adv. Phys. *24,* 645 (1975)
79. Hoare, M. R.: Adv. Chem. Phys. *40,* 49 (1979)
80. Mutaftschiev, B.: J. de Phys. Colloq. *41,* C3 (1980)
81. Echt, O., Sattler, K., Recknagel, E.: Phys. Rev. Lett. *47,* 1121 (1981)
82. Moskovits, M., Ozin, G. A.: Chryochemistry, J. Wiley, 1976, p. 395
83. Blackborrow, J. R., Young, D.: Reactivity and Structure in Organic Chemistry *9,* Metal Vapour Synthesis in Organometallic Chemistry, Springer-Verlag, Berlin, Heidelberg, New York 1979
84. Ozin, G. A., Baker, M. D., Mitchell, St. A., McIntosch, D. F.: Angew. Chem. *95,* 157 (1983), Angew. Chem. Suppl. 1983, 92; Angew. Chem. Int. Ed. Engl. *22,* 166 (1983)
85. Ozin, G. A., Mitchell, St. A.: Angew. Chem. *95,* 706 (1983), Angew. Chem. Int. Ed. Engl. *22,* 674 (1983)
86. Dietz, T. G., Duncan, M. A., Powers, D. E., Smalley, R. E.: J. Chem. Phys. *74,* 6511 (1981)
87. Kappes, M. M., Kunz, R. W., Schumacher, E.: Chem. Phys. Lett. *91,* 413 (1982)
88. Michalopoulos, D. L., Gensic, M. E., Hansen, S. G., Powers, D. E., Smalley, R. E.: J. Phys. Chem. *86,* 3914 (1982)
89. Hopkins, J. B., Langridge-Smith, P. R. R., Morse, M. D., Smalley, R. E.: J. Chem. Phys. *78,* 1627 (1983)
90. Rohlfing, E. A., Cox, D. M., Kaldor, A.: J. Phys. Chem. *88,* 4497 (1984)
91. Rohlfing, E. A., Cox, D. M., Kaldor, A.: Chem. Phys. Lett. *99,* 161 (1983)
92. Howard, J. A., Preston, K. F., Sutcliffe, R., Mile, B., Tse, J. S.: Chem. Phys. Lett. *94,* 561 (1983)
93. Howard, J. A., Preston, K. F., Sutcliffe, R., Mile, B., Tse, J. S.: J. Phys. Chem. *87,* 2268 (1983)
93b. Nazar, L. F., Ozin, G. A., Hugues, F., Godber, J., Rancourt, D.: Angew. Chem. *95,* 645 (1983), Angew. Chem. Suppl. 1983, 898; Angew. Chem. Int. Ed. Engl. *22,* 624 (1983)
94. Hoare, M. R., Pal, P.: Adv. Phys. *20,* 161 (1971)
95. Nishioka, K., Shawyer, R., Bienenstock, A., Pound, G. M.: J. Chem. Phys. *55,* 5082 (1971)
96. Burton, J. J.: ibid. *56,* 3133 (1972)
97. Briant, C. C., Burton, J. J.: Nature, Phys. Sci. *243,* 100 (1973)
98. Chini, P.: Gazz. Chim. Ital. *109,* 225 (1979)
99. Tolman, C. A.: Chem. Rev. *77,* 313 (1977)
100. Vollenbroek, F. A., Ph. D.: Thesis, University of Nijmegen 1979
101. Mingos, D. M. P.: Inorg. Chem. *21,* 464 (1982)
102. Schmid, G., Pfeil, R., Boese, R., Bandermann, F., Meyer, S., Calis, G. H. M., van der Velden, J. W. A.: Chem. Ber. *114,* 3634 (1981)
103. Cariati, F., Naldini, L.: Inorg. Chim. Acta *5,* 172 (1971)
104. van der Velden, J. W. A., Vollenbroek, F. A., Bour, J. J., Beurskens, P. T., Smits, J. M. M., Bosman, W. P.: Rec. Trav. Chim., Pay-Bas *100,* 148 (1981)
105. Vollenbroek, F. A., Bouten, P. C. P., Trooster, J. M., van den Berg, J. P., Bour, J. J.: Inorg. Chem. *17,* 1345 (1978)
106. Vollenbroek, F. A., Bour, J. J., Trooster, J. M., van der Velden, J. W. A.: J. Chem. Soc. Chem. Comm. *1978,* 907
107. Vollmert, B.: Grundriß der Makromolekularen Chemie, Vol. III, p. 43, Karlsruhe, Vollmert-Verlag 1979
108. Schmid, G., Huster, W.: unpublished results
109. Schmid, G., Giebel, U., Huster, W., Schwenk, A.: Inorg. Chim. Acta *85,* 97 (1984)
110. Johnson, B. F. G., Benfield, R. E.: in "Transition Metal Clusters", Wiley 1980, p. 471
111. Cotton, F. A., Kruczynski, L., Shapiro, B. L., Johnson, L. F.: J. Am. Chem. Soc. *94,* 6191 (1972)
112. Adams, R. D., Cotton, F. A.: Dynamic Nuclear Resonance Spectroscopy (Jackman, L. M., Cotton, F. A., Eds.), Academic Press, New York 1975, p. 520
113. Evans, J., Johnson, B. F. G., Lewis, J., Norton, J. R., Cotton, F. A.: J. Chem. Soc. Chem. Comm. *1973,* 807
114. Lawson, R. J., Shapley, J. R.: J. Am. Chem. Soc. *98,* 7433 (1976)
115. Martinengo, S., Heaton, B. T., Goodfellow, R. J., Chini, P.: J. Chem. Soc. Chem. Comm. *1977,* 39

116. Albano, V. G., Chini, P., Martinengo, S., McCaffrey, D. J. A., Strumolo, D., Heaton, B. T.: J. Am. Chem. Soc. *96,* 8106 (1974)
117. Geoffroy, G. L., Gladfelter, W. C.: ibid. *99,* 6775 (1977)
118. Gansow, O. A., Gill, D. S., Bennis, F. J., Hutchison, J. R., Vidal, J. L., Schoening, R. C.: ibid. *102,* 2449 (1980)
119. Schwenk, A.: Z. Physik *213,* 482 (1968)
120. Schwenk, A.: J. Magn. Reson. *5,* 376 (1971)
121. Schwenk, A.: private communication
122. Mingos, D. M. P.: Pure and Applied Chem. *52,* 705 (1980)
123. Wade, K.: Adv. Inorg. Chem. Radiochem. *18,* 1 (1976)
124. Lipscomb, W. N.: "Boron Hydrides", Benjamin, W. A., New York 1963
125. Gillespie, R. J.: J. Chem. Soc. Rev. *8,* 315 (1979)
125a. Briant, C. E., Hall, K. P., Wheeler, A. C., Mingos, D. M. P.: J. Chem. Soc., Chem. Commun. 1984, 248
125b. Slovokhotov, Y. L., Struchkov, Y. T.: J. Organomet. Chem. *258,* 47 (1983)
125c. Xu, G.-X.: Internat. Conf. on Coordination Chemistry, Boulder, U.S.A. 1984
126. Lauher, J. W.: J. Am. Chem. Soc. *100,* 5305 (1978)
127. Lauher, J. W.: ibid. *101,* 2604 (1979)
127a. Teo, B. K.: J. Chem. Soc., Chem. Commun. 1983, 1362
127b. Bacci, M.: Physics Letters *99A,* 230 (1983)
128. Mingos, D. M. P.: J. Chem. Soc. Dalton *1976,* 1163
129. Vollenbroek, F. A., Bour, J. J., van der Velden, J. W. A.: Recl. Trav. Chim., Pays-Bas *99,* 137 (1980)
130. van der Velden, J. W. A.: Thesis, University of Nijmegen 1983
131. Baltzer, R.: unpublished results
132. Bovin, J.-O.: unpublished results

Spectroscopic Properties of the Quasi One-Dimensional Tetracyanoplatinate(II) Compounds

Günter Gliemann and Hartmut Yersin

Institut für Physikalische und Theoretische Chemie, Universität Regensburg, D-8400 Regensburg, Federal Republic of Germany

Within the last ten years significant advances have been made in the knowledge and interpretation of the physical and chemical properties of organic and inorganic compounds with quasi one-dimensional crystal structures. Among these the $M_x[Pt(CN)_4] \cdot n\ H_2O$ compounds (with M = Na, K, Ba, Mg, Eu...) show a series of very unusual properties. These are mainly displayed by the lowest excited states which for most compounds exhibit a strong polarized luminescence. By the choice of the cation M it is possible to position the lowest transition energies within the large range from the near ultraviolet to the red. Further, high pressure application allows a continuous tuning, even down to the near infrared. By temperature variation and the application of high magnetic fields an assignment of the low-energy excited states can be achieved. Two distinctly different types of states are present, delocalized states and localized states (self-trapped excitons). By the occurrence of these different types of states the tetracyanoplatinates(II) represent interesting model systems.

Structure and Bonding 62

A. Introduction

In this review the optical properties of the tetracyanoplatinates(II), $M_x[Pt(CN)_4] \cdot n\ H_2O$ (≡MCP) are reported.

Most of the tetracyanoplatinate(II) compounds form colorful crystals, which stimulated many spectroscopic studies. Already in 1822 Gmelin[1] has observed a strong optical anisotropy of “Blausaures Platinoxydul-Kali” (potassium tetracyanoplatinate). An exact description of the polarization behavior of light reflected by crystals of the magnesium and the barium salt has been given by Brewster in 1850[2]. Encouraged by these results, in 1853 Stokes reported the “fluorescence” of several tetracyanoplatinates. He concluded: “It is only the solid state that the platinocyanides are sensitive; their solutions look like mere water”[3]. The first systematic spectroscopic studies are due to the work of Yamada on the polarized transmission of tetracyanoplatinate single crystals[4].

The interest in the optical properties of the tetracyanoplatinate compounds has increased rapidly since the detection of the highly anisotropic conductivity of the partially oxidized systems[5–12]. In the meantime it has been proved that also the non-partially oxidized tetracyanoplatinates(II) exhibit fascinating features, mainly in the optical field.

Many optical properties of the solid state are closely correlated with the quasi one-dimensional crystal structure of the tetracyanoplatinates, which is characterized by parallel columns of $[Pt(CN)_4]^{2-}$ ions. Therefrom the pronounced anisotropy of the optical transmission, reflection, and emission originates. The energies of the optical transitions sensitively depend on the Pt–Pt distance in the columns. By variation of the cations and/or by application of high pressure this distance and, therefore, the optical transition energies can be shifted over a wide range. Strong magnetic fields give rise to distinct changes of the optical emission properties (lifetimes, intensities, energies). Furthermore, the crystalline tetracyanoplatinates are appropriate model systems for investigations on energy transfer and its tunability, for example, from the $[Pt(CN)_4]^{2-}$ columns to energy sinks between or within the columns. Because of the extreme pressure dependence of the optical emission properties the tetracyanoplatinates are suited also for studies on such exotic fields as triboluminescence. Many of the properties mentioned above can be attributed to the occurrence of delocalized and of localized (self-trapped) excitons.

B. Structural Properties

The structural properties of $M_x[Pt(CN)_4] \cdot n\ H_2O$ crystals have been reported in a number of reviews[13–15]. Here we want to present a short summary on the structural aspects, which are of importance for the optical properties.

In the solid state the D_{4h} symmetry of the $[Pt(CN)_4]^{2-}$ ion is essentially maintained. The Pt–C distances vary from 1.958 to 2.026 Å, and the C–N distances from 1.119 to 1.189 Å. The angles C–Pt–C range from 87.45° to 92.16°, the angles Pt–C–N from 174.5° to 180°. The Pt atom and the four cyano groups nearly set up a plane. The deviation from the plane is not more than 0.1 Å.

In the solid state the square planar $[Pt(CN)_4]^{2-}$ complexes are stacked forming parallel columns (in most cases taken as the crystallographic c axis) with short intra-chain

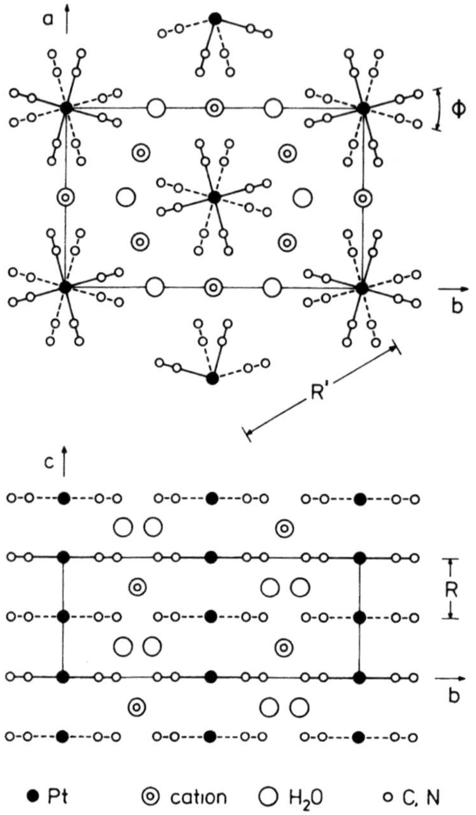

Fig. 1. Idealized structure of tetracyanoplatinates(II), projection on the *ab* and *bc* plane, respectively. *R* is the Pt–Pt distance in the chain, *R'* the distance between adjacent Pt chains and ϕ the torsion angle of adjacent complexes within the chain

Pt–Pt distances R between 3.09 and 3.75 Å (Fig. 1, Table 1). (In the partially oxidized compounds R lies between 2.798 and 2.965 Å[14, 15]. The Pt–Pt distance in Pt metal is 2.746 Å[16]).

B.I. The Single Chain

With respect to their special structural shape the Pt chains can be classified into the following groups (Fig. 2, see also Table 1):

(a) The chain is *linear* and the Pt atoms are *equidistant*. Examples are KCP · 3 H_2O, RbCP · H_2O, and BaCP · 4 H_2O.

(b) *Non-equidistant* Pt atoms in a *linear* chain are found in many partially oxidized compounds.

(c) Neighbouring Pt atoms are all *equidistant* but form a *zig zag chain*. In KNaCP · 3 H_2O the Pt atoms lie alternating ± 0.05 Å (in the direction of the *b* axis) apart from the c axis, in RbCP · 1.5 H_2O the alternating distances are -0.27, 0, and $+0.27$ Å.

Table 1. Colors and structural data of tetracyanoplatinate(II) single crystals

compound	color	chain type	R [Å]	ϕ[c] [°]	space group	chain arrangement[d]	Ref.
$Na_2[Pt(CN)_4] \cdot 3\ H_2O$	colorless	zigzag	3.651 3.754 3.691 3.745	0	$P\bar{1}$	C	17, 18
$Rb_2[Pt(CN)_4] \cdot 1\ H_2O$	colorless	linear	3.69	17	Pncn	distorted B	19
$Sr[Pt(CN)_4] \cdot 5\ H_2O$	colorless		3.60	≈0	C 2/m	B	20, 21
$Cs_2[Pt(CN)_4] \cdot 1\ H_2O$	colorless	helical	3.543	≈33	$P6_1$	C	22–24
$K_2[(Pt(CN)_4] \cdot 3\ H_2O$	light-yellow	linear	3.478	15.7–16.7	Pbcn	B	25
$Rb_2[Pt(CN)_4] \cdot 1.5\ H_2O$	pale-green	zigzag	3.421	29.5–35.2	C 2/c	B	26, 27
$Ca[Pt(CN)_4] \cdot 5\ H_2O$	yellow-green		3.38	≈30	Pcab		20, 28
$Eu_2[Pt(CN)_4]_3 \cdot 18\ H_2O$	yellow		3.37[a]				30
$Mg[Pt(CN)_4] \cdot 4.5\ H_2O$	yellow-green		3.364			B	20
$Pr_2[Pt(CN)_4]_3 \cdot 18\ H_2O$	yellow		3.36[a]				31
$Nd_2[Pt(CN)_4]_3 \cdot 18\ H_2O$	yellow		3.35[a]				31
$Ce_2[Pt(CN)_4]_3 \cdot 18\ H_2O$	yellow		3.35[a]				31
$Sm_2[Pt(CN)_4]_3 \cdot 18\ H_2O$	yellow		3.35[a]				30
$Gd_2[Pt(CN)_4]_3 \cdot 18\ H_2O$	yellow		3.35[a]				31
$K_2Sr[Pt(CN)_4]_2 \cdot 6\ H_2O$	yellow-green		3.33		$P2_1/m$		20
$Ba[Pt(CN)_4] \cdot 4\ H_2O$	yellow-green	linear	3.321	45	C 2/c	B	32, 33
$La_2[Pt(CN)_4]_3 \cdot 18\ H_2O$	yellow-green		3.32[a]				31
$Tb_2[Pt(CN)_4]_3 \cdot n\ H_2O$	yellow		3.32[a]				34
$Li_2[Pt(CN)_4] \cdot 4\ H_2O$	green		3.32[a, b]				29, 42
$(enH_2)[Pt(CN)_4] \cdot 0.5\ H_2O$	yellow-orange		3.268 ...3.286				37, 38
$LiRb[Pt(CN)_4] \cdot 2\ H_2O$	yellow		3.27[a]				34
$KNa[Pt(CN)_4] \cdot 3\ H_2O$	orange	zigzag	3.263	34.4–34.5	Cc	B	39, 40
$(NH_4)_2[Pt(CN)_4] \cdot 2\ H_2O$	pale-green		3.26				41
$Sr[Pt(CN)_4] \cdot 2\ H_2O$	orange		3.25[a]				34
$Tl_4[Pt(CN)_4](CO_3)$	red	linear	3.245		P 4/mcc	A	42
$K_2Sr[Pt(CN)_4]_2 \cdot 2\ H_2O$	orange		3.21		I		20
$KLi[Pt(CN)_4] \cdot 2\ H_2O$	light-red		3.20		$P2_1cn$		29
$Dy_2[Pt(CN)_4]_3 \cdot 21\ H_2O$	red		3.184		Pcan		43
$Tb_2[Pt(CN)_4]_3 \cdot 21\ H_2O$	red		3.181				43
$Y_2[Pt(CN)_4]_3 \cdot 21\ H_2O$	red		3.180		Pcan		43, 44
$Er_2[Pt(CN)_4]_3 \cdot 21\ H_2O$	red		3.174				43
$Ho_2[Pt(CN)_4]_3 \cdot 21\ H_2O$	red		3.17[a]				31
$Tm_2[Pt(CN)_4]_3 \cdot 21\ H_2O$	red		3.17[a]				31
$Yb_2[Pt(CN)_4]_3 \cdot 21\ H_2O$	red		3.17[a]				31
$Mg[Pt(CN)_4] \cdot 7\ H_2O$	red		3.155	45		A	20, 45
$Sr[Pt(CN)_4] \cdot 3\ H_2O$	violet-red		3.09		C	B	20
$Dy_2[Pt(CN)_4]_3 \cdot 13.5\ H_2O$	yellow	details unkown					
$Y_2[Pt(CN)_4]_3 \cdot n\ H_2O$	yellow						
$Tm_2[Pt(CN)_4]_3 \cdot n\ H_2O$	yellow						
$Yb_2[Pt(CN)_4]_3 \cdot n\ H_2O$	yellow						
$Gd_2[Pt(CN)_4]_3 \cdot n\ H_2O$	red						

[a] From spectroscopic data (see for example Ref. 29; [b] In Ref. 35 $Li_2[Pt(CN)_4] \cdot n\ H_2O$, green, is given with R = 3.17 Å; [c] See Fig. 1; [d] See Fig. 3

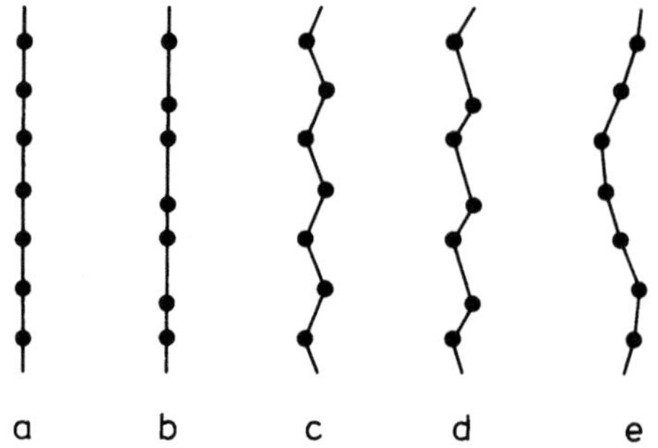

Fig. 2. Main types of individual chains (only the Pt atoms are reproduced): (*a*) equidistant linear chain; (*b*) non-equidistant linear chain; (*c*) zig zag chain; (*d*) non-equidistant zig zag chain; (*e*) helical chain[14)]

(d) Neighbouring Pt atoms are *non-equidistant* in parallel *zig zag chains*. In NaCP · 3 H_2O the Pt–Pt distances are 3.651 and 3.754 Å in one chain and 3.691 and 3.745 Å in the other crystallographically non equivalent chain. The largest deviation of Pt atoms from the c axis is 0.285 Å.

(e) The strongest deviation from a linear chain was found in CsCP · H_2O in which the Pt atoms form a *helical* chain. The equidistant Pt atoms are 1.471 Å remote from the helix axis, which is parallel to the c axis[22–24)].

The square planar $[Pt(CN)_4]^{2-}$ complex ions are not always strictly parallel to each other and not always perpendicular to the c axis. In NaCP · 3 H_2O they are parallel to each other within every chain, but form a dihedral angle of 14° with those of the next crystallographically non-equivalent chain. In KCP · 3 H_2O the normals of the complex planes deviate by ~3° from the c axis, in RbCP · H_2O ~7°, in CsCP · H_2O ~15°, in CaCP · 5 H_2O ~6°, in SrCP · 5 H_2O ~5°, in BaCP · 4 H_2O ~3°. In KNaCP · 3 H_2O the dihedral angle between two neighbouring complexes within the chain is 2.3°. In RbCP · 1.5 H_2O the planes of the complexes form alternatively dihedral angles of 8.1° and 13.7° with the ab plane.

Neighbouring complexes within a chain are rotated up to 45° around the c axis with respect to one another (Table 1). The angle ϕ may be calculated according to the minimum potential of stacked CN-ligands in dependence on the stacking distance R[47)]. Though some calculated values agree rather well with the ones observed by X-ray and neutron diffraction, other factors are also of great importance, namely the influence of the cations and the content of crystal water[14)].

B.II. Relative Ordering of the Columns in the Solids

The individual chain is separated from other chains by a distance of about 10 Å. In some compounds the complexes of one chain are shifted in the direction of the c axis with respect to the neighbouring chain. The geometrical arrangements of neighbouring chains can be approximately classified into three types, which are outlined in Fig. 3.

The chain arrangements and the intrachain Pt–Pt distances R depend on the polarizing power of the cations and on the crystal water content[14)]. The Pt–Pt distances R in the compounds of type A chain arrangement are relatively small. They lie between 3.15 and

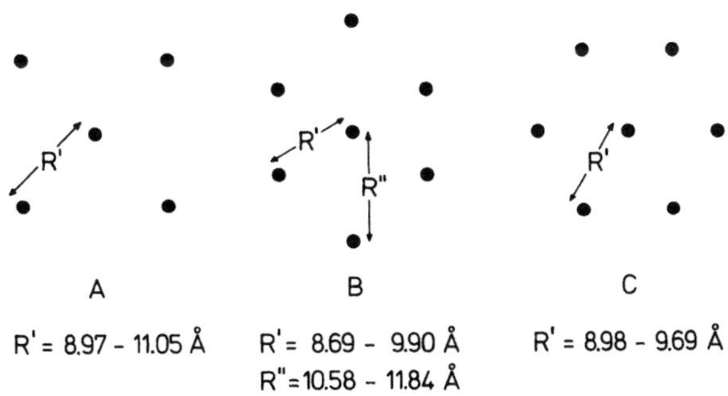

Fig. 3. Arrangement of the chains (Projection parallel to the c axis, only the Pt atoms are reproduced)[14]

3.18 Å in the red MgCP and LnCP's (and between 2.798 and 2.965 Å in the partially oxidized compounds). The tetracyanoplatinates(II) of type B chain arrangement form larger Pt–Pt distances R of 3.20–3.48 Å. The chain arrangement of type C is only found in compounds with large R, in CsCP with R = 3.54 Å and approximately in NaCP with R ≈ 3.7 Å.

B.III. Oligomeric Species

There are several indications that in concentrated aqueous solution clusters of $[Pt(CN)_4]^{2-}$ exist, probably as precursors of a chain formation [Ref. 48 and further references therein]. As shown by absorption spectroscopy the concentration of oligomeric species can be increased drastically by application of high pressure[49].

C. The Electronic Structure of the $[Pt(CN)_4]^{2-}$ Ion

The solid state properties of the linear chain compounds are substantially determined by the electronic structure of the molecular $[Pt(CN)_4]^{2-}$ units. An ab initio calculation of the electronic structure of the $[Pt(CN)_4]^{2-}$ complex including all 132 electrons or at least the 48 valence electrons does not exist as yet. Concerning the optical spectroscopy, however, mainly the highest occupied (HOMO) and the lowest unoccupied (LUMO) molecular orbitals are of interest. Thus, the number of electrons and states, which have to be considered, is restricted drastically compared to the full problem. The method usually applied to the approximative determination of the relevant states and their electronic structures is the molecular-orbital ligand field theory mostly using empirical fitting data[50, 51].

A decision which orbitals of the central ion and of the ligands determine the HOMO's and the LUMO's is not trivial and can be based on theoretical estimates and/or experimental information. We want to discuss this subject in some detail:

The highest occupied one-electron Pt(II) orbitals are certainly of 5 d character[52]. From a ligand field approach for the $[Pt(CN)_4]^{2-}$ complex with D_{4h} symmetry (cf. Sect. B) follows that the orbitals $5d_{xy}(b_{2g})$, $5d_{xz},d_{yz}(e_g)$, $5d_{z^2}(a_{1g})$ belong to the ground state configuration, while $5d_{x^2-y^2}(b_{1g})$ is destabilized by more than 30,000 cm^{-1}, compared to the others[53]. The energetical sequence of the four occupied 5d orbitals, however, is still controversial[48, 53–67].

In order to understand the optical properties one must know which orbitals of the central ion and of the ligands compose the LUMO's. Beside the $5d_{x^2-y^2}$ orbital one has also to consider the Pt $6p_z$ and Pt 6s orbitals, since the energies of the corresponding states of the free Pt^{2+} ion lie in an optically accessible energy range[52]. As the molar extinction coefficients of the low-energy absorption bands are very large ($\varepsilon \approx 10^4$ l/mol cm, see Fig. 4), the corresponding transitions will be symmetry allowed. Consequently, the parity-forbidden d–d and d–s transitions are ruled out. Only the $6p_z$ orbital has the appropriate ungerade symmetry.

The CNπ^* ligand orbitals of symmetry a_{2u} can also be excited by a parity allowed process. It is agreed throughout the literature that the strong excitations of the tetracyanoplatinate complexes in the near ultraviolet have pronounced metal-to-ligand charge transfer (MLCT) character and are connected with the transition Pt 5 d $\rightarrow$ CNπ^*. The MLCT character is concluded from a comparison of $[Pt^{II}(CN)_4]^{2-}$ absorption spectra with those of $[Au^{III}(CN)_4]^-$. Au(III) and Pt(II) have the same groundstate electron configurations, but the highest occupied orbitals of Au(III) are more strongly stabilized by the larger oxidation number. Consequently, MLCT transitions of the gold complexes are

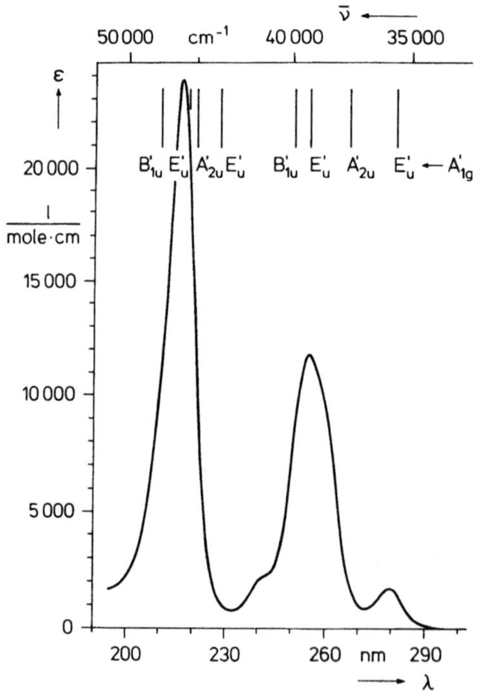

Fig. 4. Absorption spectrum of $[Pt(CN)_4]^{2-}$ (low concentration) in aqueous solution at 295 K[68]. The double group states calculated by Isci and Mason[62] are reproduced for comparison. Energy states without spin-orbit coupling are used as input parameters: (in 10^3 cm^{-1}) $^1B_{1u}$: 47.0; $^3B_{1u}$: 43.0; ^1E_u: 43.5; ^3E_u: 40.5; $^1A_{2u}$: 42.0; $^3A_{2u}$: 37.5, and ζ_{5d} (complex): 3.0

expected to be shifted to higher and LMCT transitions to lower energies, respectively. Indeed, the CT-transitions of the $[Au(CN)_4]^-$ complexes are observed at about 10,000 cm^{-1} higher energies than those of the platinum complexes[53].

Thus, the molecular orbital (of ungerade symmetry a_{2u}) representing the lowest excited one-electron state will be mainly composed of CNπ^* and Pt 6p_z wavefunctions[56,57]. The assumption of a_{2u} as the mainly relevant excited MO leads to a satisfactory interpretation of the spectra (see below and the references[58,62,67]). MO calculations place the next excited one-electron state of ungerade symmetry (e_u) distinctly higher in energy ($\Delta \approx 16{,}000$ cm^{-1}, see the references[59,60]), which allows to disregard this state in the limits of the actual approach.

The above arguments are summarized in Fig. 5. The low energy excitations correspond to the transitions (a_{1g}, e_g, b_{2g}) $\rightarrow$ a_{2u}. Interrante and Messmer[59,60] have calculated the spatial extent of these one-electron MO-functions using the SCF-Xα-SW theory (self-consistent-field Xα scattering-wave)[69,70]. For these calculations they divided the space into three regions. Region I consists of spherical volumes located around the centers of the respective atoms, being tangent to one another (or slightly overlapping). Region III is the space outside a sphere which has its origin at the center of the molecule and includes all atomic spheres of region I, and which is tangent to these. Finally, region II is the space interstitial to the spheres of region I and III. The Xα equations are solved for the three regions, using three different exchange parameters α which result from atomic calculations[71,72]. For further details see the refs. 59, 60, 69, 70, 73.

Results are given in Table 2 and in the Figs. 6 and 7. Table 2 summarizes the spatial charge distribution of the molecular orbitals mentioned above. This approach shows as an important feature the extensive mixing of metal and ligand orbitals indicating the

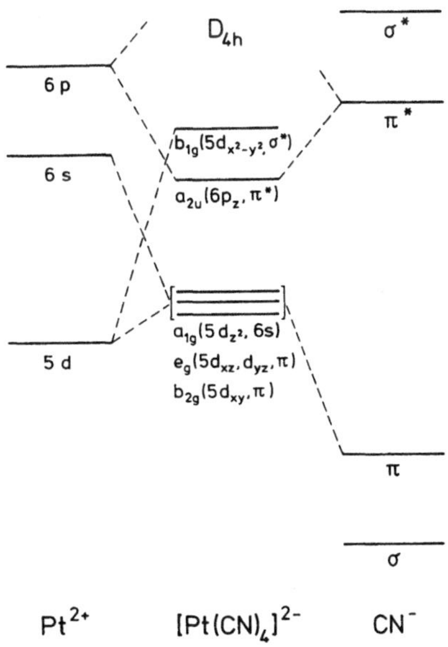

Fig. 5. Simplified one-electron MO diagram for $[Pt(CN)_4]^{2-}$. The lowest unoccupied orbital a_{2u} is strongly stabilized by an interaction of Pt 6p_z and CNπ^* states. The sequence of the highest occupied orbitals cannot be given. Electric dipole transitions in D_{4h} are only allowed for $a_{1g} \leftrightarrow a_{2u}$ and $e_g \leftrightarrow a_{2u}$

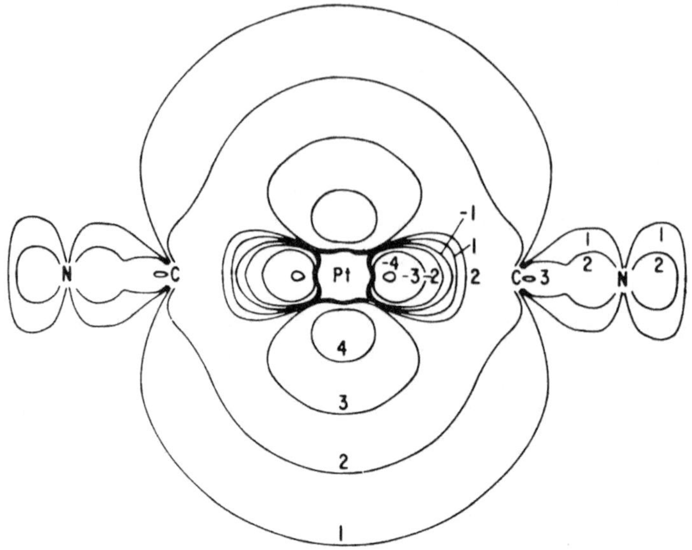

Fig. 6. Contour plot of the a_{1g} MO wavefunction resulting from SCF-Xα-SW calculations. The amplitude of the wavefunction changes from one plot to the other by a factor of three. Intermolecular distances: Pt–C = 2.02 Å, C–N = 1.11 Å[60]

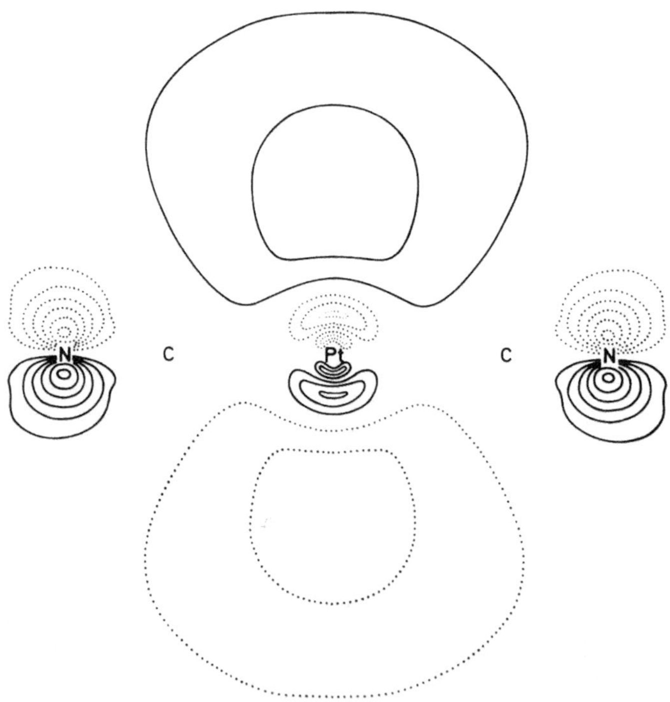

Fig. 7. Contour plot of the a_{2u} MO. The values of the wavefunction amplitudes are not given[61]

Table 2. MO-charge distribution for $[Pt(CN)_4]^{2-}$. The r-values determine the spatial extent of the atomic spheres. These and the regions are defined in the text[60, 61]

Orbital	Percentage of charge distribution				
	region I (atomic spheres)			region II	region III (outer-sphere)
	Pt $r = 1.44$ Å	4 C $r = 0.75$ Å	4 N $r = 0.69$ Å		$r = 3.8$ Å
$a_{2u}(6p_z, CN\pi^*)$	6.4	0.1	10.4	71	12.1
$a_{1g}(5d_{z^2}, 6s)$	72.7	4.2	1.4	21.3	0.4
$e_g(5d_{xz}, d_{yz}, CN\pi)$	25.9	14.4	31.1	26.9	1.8
$b_{2g}(5d_{xy}, CN\pi)$	34.0	10.8	28.4	24.7	2.0

appreciable covalency of the metal-ligand bond. For example, the charge of the e_g orbital is confined only to $\approx 26\%$ to the inner Pt sphere but about 45% (14.4% + 31.1%) are found in the CN spheres. Only the a_{1g} orbital has mainly metal character ($\approx 73\%$). The Figs. 6 and 7 reproducing the contour-curves for the a_{1g} and a_{2u} MO's demonstrate also the covalency of the electronic structure. The a_{2u} MO extends considerably into the space, increasingly resembling a Pt 6 p_z orbital. The Xα calculation for the relative charge distribution of a_{2u} in region III (radius 3.8 Å) yields a value of more than 12%. This is essential for the solid state behavior of the tetracyanoplatinates. Since in the solid state the intrachain metal-metal distances range from 3.09 Å to 3.7 Å, a very strong overlap between the wavefunctions of neighbouring $[Pt(CN)_4]^{2-}$ complex ions is expected. Further the a_{1g} MO contains an appreciable Pt 6 s admixture[60].

Interrante and Messmer[59, 60] and also Gray and Ballhausen[53] give the following energetical order of one-electron states:

$$e_u(\mathrm{Pt}\,6p, CN\pi^*) \gg b_{1g}(\mathrm{Pt}\,5d_{x^2-y^2}) \gg a_{2u}(\mathrm{Pt}\,6p_z, CN\pi^*) \gg$$

$$b_{2g}(\mathrm{Pt}\,5d_{xy}, CN\pi) > e_g(\mathrm{Pt}\,5d_{xz}d_{yz}, CN\pi) > a_{1g}(\mathrm{Pt}\,5d_{z^2}, 6s)\ . \qquad (1)$$

This result differs in the sequence found by Piepho, Schatz, and McCaffery[58] and by Isci and Mason[62].

A more recent Xα-calculation by Lopez, Yang and Case[74] including the strong spin-orbit coupling ($\zeta_{Pt} \approx 4000\ \mathrm{cm}^{-1}$)[58] leads to a substantial mixing of the states a_{1g}, e_g, b_{2g} and thus modifies the non-relativistic results of Interrante and Messmer. For example, through spin-orbit coupling the b_{2g} orbital gets a charge distribution of 27% Pt 5 d_{xy} (α spin) and 32% Pt 5 $d_{xz}d_{yz}$ (β spin) along with 41% ligand character. The charge distribution of the a_{1g} MO is also modified. The relativistic effects give rise to a charge contribution of 32% ligand character compared to about 10% in the non-relativistic case. Lopez et al.[74] apply their calculations also to trimers of $[Pt(CN)_4]^{2-}$ and extend their discussions to linear chains.

Taking the electron-electron interaction into account the ground state configuration $a_{1g}^2 b_{2g}^2 e_g^4$ yields the term $^1A_{1g}$ whereas each of the excited configurations, resulting from one-electron transitions between the ground state configuration and the lowest unoccupied a_{2u} MO, forms a singlet and a triplet, as shown in Table 3. Spin orbit coupling

Table 3. Irreducible representations of the lowest excited electronic states of $[Pt(CN)_4]^{2-}$ resulting from excitations to the lowest excited a_{2u} molecular orbital. Electric dipole transitions are only allowed from the ground state A'_{1g} to $A'_{2u}(z)$ and $E'_u(x, y)$, respectively

Configuration	Many-electron states in D_{4h} without spin-orbit coupling	Double group states in D'_{4h} with spin-orbit coupling
$b_{2g}^2e_g^4a_{1g}^2a_{2u}^0$ (ground state)	$^1A_{1g}$	A'_{1g}
$b_{2g}^2e_g^4a_{1g}^1a_{2u}^1$	$^1A_{2u}$	A'_{2u}
	$^3A_{2u}$	A'_{1u}, E'_u
$b_{2g}^2e_g^3a_{1g}^2a_{2u}^1$	^1E_u	E'_u
	^3E_u	A'_{1u}, A'_{2u}, B'_{1u}, B'_{2u}, E'_u
$b_{2g}^1e_g^4a_{1g}^2a_{2u}^1$	$^1B_{1u}$	B'_{1u}
	$^3B_{1u}$	B'_{2u}, E'_u

splits the triplet terms and mixes states of the same representation. The wavefunctions $|A'_{2u}i\rangle$ and $|E'_ui\rangle$ of the double group states in D'_{4h} symmetry have the following forms:

$$
\begin{aligned}
|A'_{2u}i\rangle &= a_{1i}|A'_{2u}(^1A_{2u})\rangle + a_{2i}|A'_{2u}(^3E_u)\rangle \quad \text{with } i = 1, 2 \\
|E'_ui\rangle &= b_{1i}|E'_u(^3B_{1u})\rangle + b_{2i}|E'_u(^3A_{2u})\rangle \\
&\quad + b_{3i}|E'_u(^3E_u)\rangle + b_{4i}|E'_u(^1E_u)\rangle \quad \text{with } i = 1, 2, 3, 4\,.
\end{aligned} \tag{2}
$$

Piepho, Schatz, and McCaffery[58] and Isci and Mason[62] have determined the coefficients a_{ji} and b_{ji} by diagonalizing the spin-orbit coupling matrices. In these investigations estimated energies for the unperturbed singlet and triplet states and for the spin-orbit parameter ζ were used. The input parameters were mainly determined from (and fitted to) the optical absorption spectra of $[Pt(CN)_4]^{2-}$ solutions in consistency with the MCD spectra. The spin-orbit matrix elements were calculated by approximating the occupied MO's as pure d functions. The a_{2u} orbital was assumed to consist of carbon 2p functions exclusively.

The mixing coefficients of the wavefunctions for the double group states A'_{2u} and E'_u obtained by Isci and Mason[62] are shown in Table 4. (The values given by Piepho et al.[58] are slightly different.) For example, the wavefunctions $|A'_{2u}1\rangle$ and $|A'_{2u}2\rangle$ have the following forms

$$
\begin{aligned}
|A'_{2u}1\rangle &= -0.632|A'_{2u}(^1A_{2u})\rangle + 0.774|A'_{2u}(^3E_u)\rangle\,, \\
|A'_{2u}2\rangle &= 0.774|A'_{2u}(^1A_{2u})\rangle + 0.632|A'_{2u}(^3E_u)\rangle\,.
\end{aligned} \tag{3}
$$

The mixing coefficients of the parentage wavefunctions are nearly equal in both double group states, which implies, that transitions to $A'_{2u}1$ and $A'_{2u}2$ should be observed with nearly the same oscillator strengths. This is in agreement with the polarized absorption measurements on single crystals. The absorption spectra show two transitions, which have the corresponding polarization behavior and exhibit nearly the same molar extinction coefficients (see Fig. 8). Although, $^1A_{2u}$ and ^3E_u belong to different one-electron

Table 4. Eigenstates and mixing coefficients for the double group states A'_{2u} and E'_u for $[Pt(CN)_4]^{2-}$. The data are taken from Isci and Mason (Ref. 62, suppl. material). For input parameters see the Fig. 4

	A'_{2u}-eigenstates		
	energy [cm^{-1}]	$^1A_{2u}$	^3E_u
$A'_{2u}(1)$	37,500	−0.632	0.774
$A'_{2u}(2)$	45,000	0.774	0.632

	E'_u-eigenstates				
	energy [cm^{-1}]	$^3B_{1u}$	$^3A_{2u}$	^3E_u	^1E_u
$E'_u(1)$	35,600	−0.045	0.882	−0.424	0.202
$E'_u(2)$	39,200	0.468	0.226	0.672	0.528
$E'_u(3)$	43,900	0.843	−0.192	−0.502	−0.025
$E'_u(4)$	45,700	−0.263	−0.367	−0.341	0.824

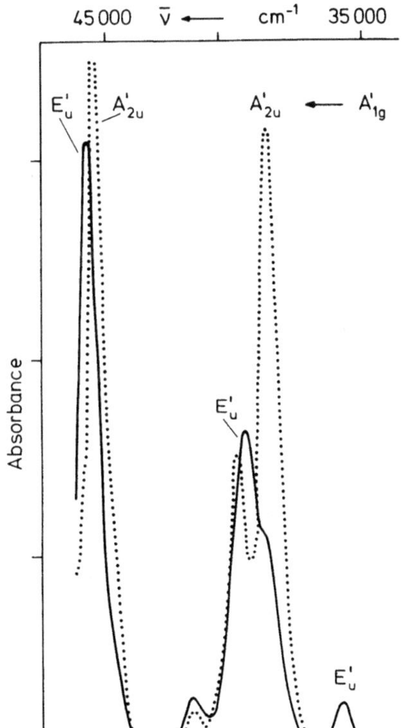

Fig. 8. Polarized absorption spectrum of a single crystal $(n\text{-}Bu_4N)_2[Pt(CN)_4]$ at 5 K. Solid line: $\vec{E}\perp\vec{C}_4$ spectrum (σ); dotted line: $\vec{E}\|\vec{C}_4$ spectrum (π) ($\vec{E}$ = electric field vector, $\vec{C}_4$ = fourfold axis of the $[Pt(CN)_4]^{2-}$ complex). The assignment to double group states results from group-theoretical selection rules; $(n\text{-}Bu_4N)$ = tetra-n-butylammonium[67]

configurations the spin-orbit coupling strongly mixes these configurations confirming the results of the above mentioned relativistic Xα calculations[74].

Isci and Mason calculated the eigenvalues of eight double group states. The energy values are compared in Fig. 4 with the absorption spectrum of an aqueous solution of $[Pt(CN)_4]^{2-}$. The applied input parameters are given below the figure.

The assignment of double group states to the experimental transition energies has been investigated also by other authors[53, 55–62]. There is complete agreement concerning the totally symmetric ground state $A'_{1g}(^1A_{1g})$. The assignments of the excited states, however, differ. They are summarized in 1976 in short papers by Miller and Marsh[65, 66].

A method directly informing about the symmetry character of the excited states has been applied by Cowman and Gray[67]. They measured the polarized absorption spectra of $(n\text{-}Bu_4N)_2[Pt(CN)_4]$ single crystals. The $[Pt(CN)_4]^{2-}$ units in the solid state are assumed to have D_{4h} symmetry. They are fairly well isolated because of the bulky cations and thus the metal-metal interaction can be ignored. The transition of lowest energy appears at 35,520 cm^{-1} for the crystal and at 35,400 cm^{-1} for the complex in solution. Cowman and Gray measured the absorption spectra with the polarization parallel and perpendicular to a crystal extinction direction, respectively. Subsequently they mathematically transformed them into the polarization components parallel (π) and perpendicular (σ) to the C_4 axis of the complex, respectively. The transformed spectra are reproduced in Fig. 8. From group theoretical selection rules ($A'_{1g} \leftrightarrow A'_{2u}$ allowed for $\vec{E}\|\vec{C}_4$; $A'_{1g} \leftrightarrow E'_u$ allowed for $\vec{E}\perp\vec{C}_4$; $\vec{E}$ = electric dipole vector) one can classify directly at least five observed transitions as shown by the symmetry lables at the peaks of the absorption spectra. This assignment confirms largely that given by Isci and Mason[62]. The fact of nearly equal values of the molar extinction coefficients for the two $A'_{1g} \rightarrow A'_{2u}$ transitions, which has already been discussed, represents a further support of the proposed interpretation.

Cowman and Gray[67] also specify the parentage configurations, which yield the main contributions to the double group wavefunctions. Schindler, Fukuda and Adamson[48], interpreting the concentration dependence of the absorption spectra of aqueous solutions of $K_2[Pt(CN)_4]$ and $Ba[Pt(CN)_4]$, come to a different result for the mixing contributions to the lowest excited states. However, the general problems concerning this assignment have already been mentioned above.

Summarizing the results of this section, it seems to be well established that the near UV absorption spectra of $[Pt(CN)_4]^{2-}$ complexes are mainly determined by transitions from occupied MO's of Pt 5dCNπ character to the unoccupied MO of Pt 6p_zCNπ^* character. The spin-orbit coupling proves to be strong. Thus, the group-theoretical assignment of the excited states is successfully accomplished within the relevant D'_{4h} double group. The electric dipole allowed transitions from the A'_{1g} ground state to three (of the four) excited E'_u and two A'_{2u} states seem to be identified by experimental methods and model calculations. The transitions to the A'_{1u}, B'_{1u} and B'_{2u} states and to states involving the $5d_{x^2-y^2}$ orbital, which are electric-dipole forbidden but allowed by coupling to adequate vibrations are still unidentified for the $[Pt(CN)_4]^{2-}$ complex ion.

D. The Quasi One-Dimensional $[Pt(CN)_4]^{2-}$ Chain

D.I. One-Electron Band Structure

Model calculations for $[Pt(CN)_4]^{2-}$-dimers by Interrante and Messmer[60] yielded a very large splitting of the Pt 5 d_{z^2} and Pt 6 p_z dimer states. Further, these calculations clearly show that the highest occupied and the lowest unoccupied states of the $[Pt(CN)_4]^{2-}$ *chain* have to be constructed mainly from Pt 5 d_{z^2} and Pt 6 p_z complex states, respectively. The dimer calculations allow an estimate of the electronic band widths (≈ twice the splitting by dimerization) of several 10^4 cm^{-1}. Therefore[75–77], one-electron band structure calculations for the tetracyanoplatinates will be appropriate. Results of such calculations have been reported by several groups[78–86]. Especially the calculation of the band structure for the partially oxidized KCP compound by Messmer and Salahub[81] stimulated the discussion. The authors used Pt(5 d_{z^2},6 s) hybrids and Pt 6 p_z orbitals as basis functions to construct the Bloch functions. This type of basis set turned out to be also appropriate for an approximative description of optical properties of the tetracyanoplatinates(II)[82].

These calculations start with the Bloch functions[75–77] for a chain:

$$\chi_j(k, r) = N^{-1/2} \sum_{m=1}^{N} e^{ikmR} \phi_j(r - mR). \quad j = 1, 2 . \tag{4}$$

R is the intrachain separation between neighbouring $[Pt(CN)_4]^{2-}$ complexes. k is the absolute value of the wavevector in chain direction restricted to the first Brillouin zone $(-\pi/R < k \leq +\pi/R)$. The number of complexes in the chain is assumed to be N. Periodic boundary conditions are used.

The $\phi_j(r - mR)$, $j = 1, 2$, are the above mentioned basis functions obtained from the $[Pt(CN)_4]^{2-}$ complex ion centered at $m \cdot R$ in the chain. The wavefunction ϕ_1 for the hybrid Pt(5 d_{z^2}, 6 s) can be approximated – in analogy to Ref. 81 – by a constant mixture of atomic Slater orbitals (90% Pt 5 d_{z^2}, 10% Pt 6 s), whereas a pure Pt 6 p_z Slater orbital is used for the basis function ϕ_2.

From extended Hückel calculations the band structure of the system including band-band coupling is derived straightforwardly. Using the following approximation[87, 88] for the non-diagonal matrix elements of the Hamiltonian

$$H_{ij} \approx \tfrac{1}{2}(I_i + I_j)(2 - |S_{ij}|) \cdot S_{ij} \tag{5}$$

(S_{ij}: overlap integrals; I_i, I_j: ionization potentials, values from the SCF-Xα-SW calculations of Ref. 60) the structures of the valence and the conduction bands are obtained for different R values as shown in the upper part of Fig. 9.

In Fig. 10 the calculated band gap energies E_G are compared with the experimental emission and reflectivity energies of the $\vec{E}\|\vec{c}$-polarized transitions (see Sect. E.I.) which are correlated in the limit of the one-electron approximation. Apparently the one-electron model calculations can be used to fit the experimentally observed trends. A similar fit has been obtained in Ref. 83. Further investigations concerning the one-electron band structure calculations are found in Ref. 84 and in a very recent paper[85].

A further aspect is worthwhile to be mentioned: The lower part of Fig. 9 gives the percentage of admixture of the orbitals of a_{2u} parentage to the valence band orbitals. This

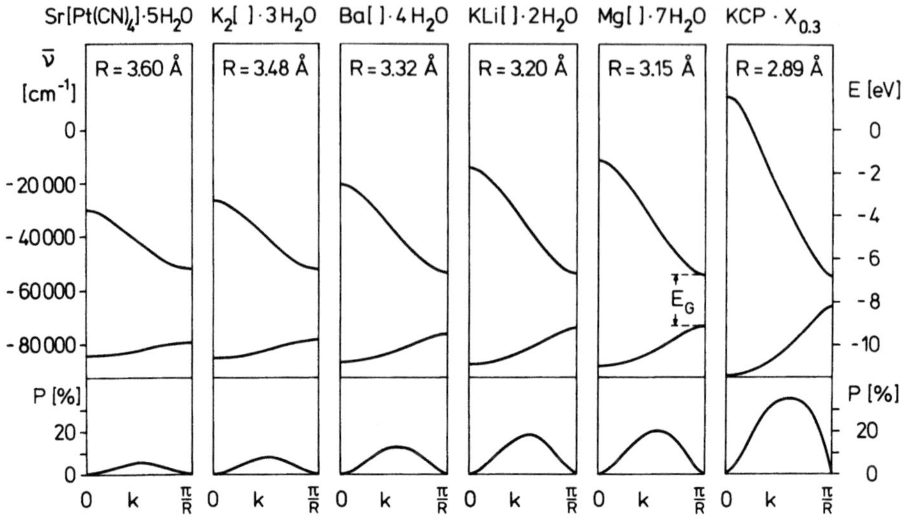

Fig. 9. Energy bands of tetracyanoplatinates with different intrachain Pt–Pt distances R. The results are obtained from a two-band model in the limit of an one-electron calculation[82]. The upper and lower structures represent the conduction and the valence bands, respectively. k is the absolute value of the wavevector in the chain direction. P is the percentage of admixture of the orbitals of a_{2u} parentage to the valence band orbitals

admixture increases with decreasing Pt–Pt distance R. The band structure calculations do not explicitely take into account the delocalization of the molecular orbital over the whole $[Pt(CN)_4]^{2-}$ complex by the strong mixing of the a_{2u} orbitals Pt 6 p_z and CNπ^* (see Sect. C.). Nevertheless, the MO character is included indirectly through the values of the ionization potentials which were taken from the MO-calculations. Thus, one can con-

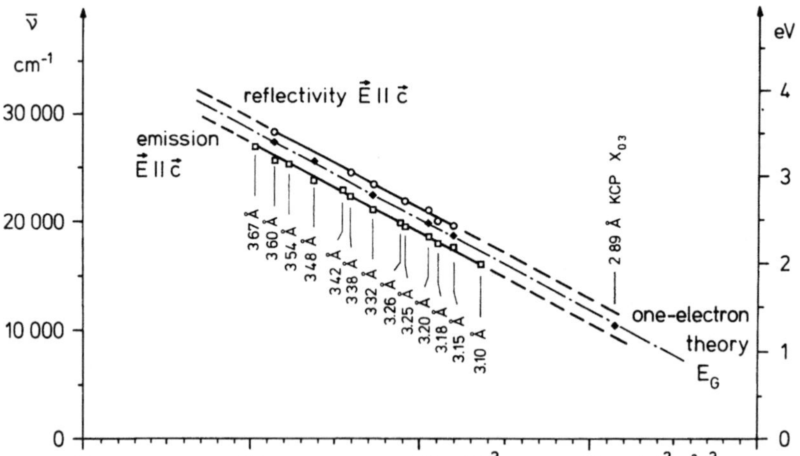

Fig. 10. Emission and reflectivity peak energies for the $\vec{E}\|\vec{c}$ polarized transitions (T = 295 K, see Table 5 and Fig. 18) of various $M_x[Pt(CN)_4] \cdot n\ H_2O$ compounds compared to the calculated band gap energies E_G (see Fig. 9)[82]

clude[82] from Fig. 9 that for decreasing R the valence band states are increasingly delocalized over the whole planar complex unit. Hence, the ground state electron charge density at the CN-ligands increases with R reduction. From this fact a weakening of the CN-stretching vibration is expected (see also Ref. 89). Moreover, the coupling between the chain and its surroundings will be strengthened, in particular for systems with short Pt–Pt distances as found in the partially oxidized tetracyanoplatinates.

As will be shown in the following sections the results of the one-electron band structure calculations allow to describe several important properties of the tetracyanoplatinates, like the dominance of the Pt 5 d_{z^2} and Pt 6 p_z orbitals for the red-shift of the main optical transitions with decreasing metal-metal distance[82–86], the admixture of Pt(6 p_z, CNπ^*) character into the valence band[82, 89], or several stabilization effects upon partial oxidation[84]. On the other hand, a series of experimentally found features is out of the scope of the one-electron band model. In the following some of these properties are specified:

- The emission behavior of the tetracyanoplatinates(II) is obviously determined also by *localized* states. Several independent investigations lead to the result that the states are localized at a $[Pt(CN)_4]^{2-}$ complex or at a cluster of only a few complexes:
  1) A small $[Ni(CN)_4]^{2-}$ impurity concentration selectively quenches the $\vec{E}\perp\vec{c}$ polarized emission. The temperature behavior of the quenching properties is explained by a phonon-assisted hopping from localized states of the chain to the $[Ni(CN)_4]^{2-}$ quencher[90–92].
  2) The $\vec{E}\perp\vec{c}$ polarized emission lifetime is reduced at low temperatures in the presence of rare-earth impurities. One observes a non-exponential decay. The long-living component equals the intrinsic one (without quencher) and the short-living component is attributed to regions emitting near the impurities (Ref. 93, see also Sect. G.).
  3) A blue-shift in the $\vec{E}\perp\vec{c}$ polarized emission under high magnetic fields corresponding to the energy of a vibration of the $[Pt(CN)_4]^{2-}$ ion, is also explained by assuming localized states (Sect. G., Ref. 94, 95).
  4) An exchange of the crystal water by D_2O strongly increases the $\vec{E}\perp\vec{c}$ polarized emission intensity. This behavior is expected for localized states coupled to the crystal water[68].
  5) An energy barrier occurs between the emitting states (Sect. E.II.1, Fig. 26).

- The intrinsic photoconductivity of BaCP crystals sets in at a photon energy of about $4 \cdot 10^3$ cm^{-1} higher than the low-energy flanc of the $\vec{E}\parallel\vec{c}$ polarized main absorption peak[96].

Two complementary models can be developed to describe the features mentioned above. Both models consider the electron-electron interaction and the exciton-lattice interaction. Although both models have different starting points, they yield satisfying interpretations of the experimental findings. The one model is the model of Wannier excitons which ensues directly from the one-electron band model discussed above. The other one starts from the many electron states of the $[Pt(CN)_4]^{2-}$ ion and takes into account the coupling between neighbouring complex ions.

D.II. Free and Localized Excitons

If an electron is excited from the valence band into the conduction band (band gap energy E_G) it leaves a positive hole in the valence band[75–77]. By the Coulomb attraction between hole and electron a stabilization results leading to free excitons FE (binding energy E_B). This stabilization relative to the energy of the dissociated FE, which corresponds to the exciton continuum, depends on the mean distance between electron and hole. Commonly, the transitions between the ground state and the FE states are found in absorption and/or emission. E_G can be determined from the photoconductivity. (At $h\nu \geq E_G$ the exciton dissociates and the onset of an intrinsic photocurrent can be observed.)

By the creation of an exciton the charge distribution is changed, and the lattice can react by a stabilizing deformation which is bound to the exciton[97–101]. If this deformation energy is relatively small it results an excitonic polaron, in which the deformation moves together with the exciton. In the case of a relatively large stabilization the exciton may cease moving and, thus, digs its own trap. A self-trapped exciton (STE) is created[99–101]. Under suitable conditions both, excitonic polarons and STE states, can be observed[101–105].

For the tetracyanoplatinates(II) it is assumed, as shown below, that the STE is restricted to a region of a few $[Pt(CN)_4]^{2-}$ units with a reduced intrachain metal-metal distance. An analogous behavior with respect to the shrinkage of the Pt–Pt distance in the excited state has been found recently to occur in the binuclear $[Pt_2(H_2P_2O_5)_4]^{4-}$ compound (shrinkage $\approx 7\%$)[106, 107].

Starting with the one-electron band structure calculation (see Sect. D.I.) the FE binding energies E_B are determined in the model of an anisotropic continuum[108, 109]. The anisotropic input parameters (dielectric constants ε_0, ε_∞; effective electron mass m_e, hole mass m_h) depend on R for the intrachain direction and are treated as constants for the transverse direction. Utilizing values which reasonably describe the physical properties of the tetracyanoplatinates, the following binding energies are obtained: E_B (R = 3.15 Å) = $2.5 \cdot 10^3$ cm^{-1}; E_B (R = 3.4 Å) = $4.7 \cdot 10^3$ cm^{-1}; E_B (R = 3.6 Å) = $6.4 \cdot 10^3$ cm^{-1}[110, 111]. These calculations have been carried out to reproduce the exciton binding energy of BaCP (R = 3.32 Å) which has been determined experimentally[96] to $\approx 4 \cdot 10^3$ cm^{-1}. (The binding energy E_B becomes at R = 3.6 Å comparable to the width of the valence band which indicates a limit of the applied effective mass approximation.)

The theoretical formalism for a description of the STE stabilization energy relative to the FE state represents a generalization of Toyozawa's model[99, 112], considering the effects of electron-hole and electron-phonon interaction in an elastic and dielectric continuum of high anisotropy. For tetracyanoplatinates the total energy of the FE or the STE can be approximated by the following expression[109–111]:

$$E(\alpha_\perp, \alpha_\parallel) = \frac{\hbar^2 \pi}{2M_\perp} \cdot \left[2\frac{\alpha_\perp^2}{a^2} + \frac{M_\perp}{M_\parallel} \cdot \frac{\alpha_\parallel^2}{R^2} \right] - \frac{1}{2} \frac{D_\parallel^2}{c_{33}} \cdot \frac{\alpha_\perp^2}{a^2} \cdot \frac{\alpha_\parallel}{R}, \qquad (6)$$

where $M_{\perp,\parallel}$ is the anisotropic mass of the exciton, R is the intrachain Pt–Pt distance and a the lattice constant perpendicular to the chain axis, $D_\parallel = \Delta E_G/(\Delta R/R)$ is the deformation potential and c_{33} the elastic constant, determined from compressibility data[113–115]. The parameters $\alpha_\perp$ and $\alpha_\parallel$ characterize the extension of the exciton-envelope perpendicular and parallel to the chain, respectively. $\alpha_{\parallel} = 0$ means the exciton is free and delocalized along the whole chain while $\alpha_{\parallel} \approx 1$ corresponds to a strongly localized exciton envelope, representing a STE.

The total energy of the exciton $E(\alpha_{\perp}, \alpha_{\parallel})$ has been calculated for MgCP[111] for the R-range from 3.15 Å to ≈3.0 Å, corresponding to a pressure range from 0 to 20 kbar[115] (see also Sect. F.I.). A series of contour maps $E(\alpha_{\perp}, \alpha_{\parallel}, R)$ results. These can be evaluated further by introducing a modified localization parameter $\tilde{\alpha}(\alpha_{\perp}, \alpha_{\parallel})$[111], which is chosen as trajectory of lowest energy on the $E(\alpha_{\perp}, \alpha_{\parallel})$ hyper-surface with R as parameter. The $E(\tilde{\alpha})$ curves are reproduced for five different R values in Fig. 11. For R > 2.95 Å the total energy $E(\tilde{\alpha})$ shows two types of minima. The minimum at $\tilde{\alpha} = 0$ (normalized to zero for the different R values) represents the free exciton (FE) state. The other minimum, separated by a potential barrier from the first one, is found near $\tilde{\alpha} = 1$ which formally corresponds to a strongly localized exciton wavefunction. The extension of the exciton envelope equals to about one lattice constant. At this $\tilde{\alpha}$ value, however, the limits of the validity of the continuum model is reached.

As shown in Fig. 12 the energy difference between the FE (at $\tilde{\alpha} = 0$) and the STE states (at $\tilde{\alpha} = 0.8$) is in quantitative agreement with the observed energy difference of the $\vec{E}\parallel\vec{c}$ and the $\vec{E}\perp\vec{c}$ polarized emission. Figure 12 shows further that the calculated energies $E_G - E_B$ of the FE states also agree with the experimental data. A comparison of the calculated band gap energies E_G is not yet possible, since the corresponding experimental data, like photoconductivity or electro-reflection[116], are not available for MgCP.

With decreasing R the minimum designating the STE becomes destabilized and presumably ceases to exist for R values below about 3 Å (Fig. 11). This is in accordance with

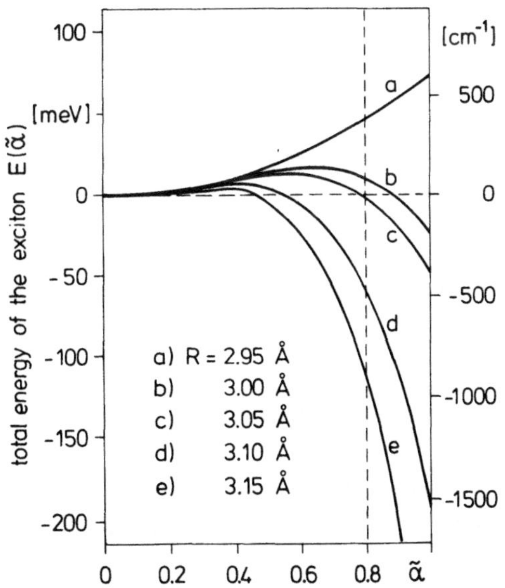

Fig. 11. Total energy of an exciton in an anisotropic elastic continuum for different Pt–Pt distances R[111]. The energy is calculated for $Mg[Pt(CN)_4] \cdot 7\, H_2O$ from Eq. (6). $\tilde{\alpha}(\alpha_{\perp}, \alpha_{\parallel})$ represents a localization parameter which describes a free exciton (FE) with $\tilde{\alpha} = 0$ and a localized exciton (self-trapped exciton STE) with $\tilde{\alpha} \approx 1$. The exciton binding energy E_B is normalized to zero for different R-values

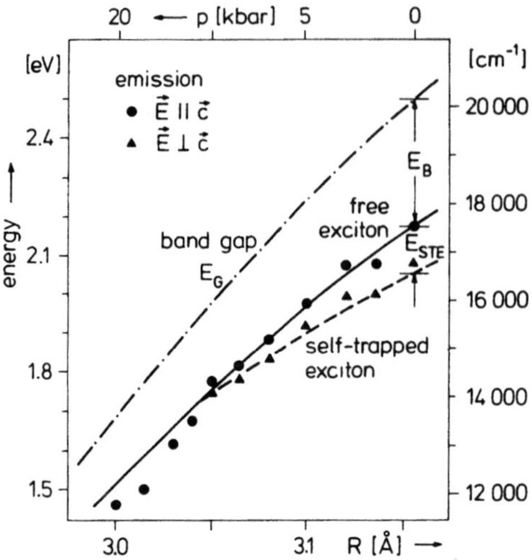

Fig. 12. Comparison of the calculated[111] results for the band gap energy $E_G(R)$, the free exciton energy $E_G(R) - E_B(R)$ and the self-trapped excition energy $E_G(R) - E_B(R) - E_{STE}$ ($\tilde{\alpha} = 0.8$) with the differently polarized emission peak energies (from Ref. 113–115) for $Mg[Pt(CN)_4] \cdot 7\ H_2O$

the vanishing of the $\vec{E} \perp \vec{c}$ emission for small R values (high pressures). Moreover, from the model calculations a potential barrier results between the FE and STE states, probably corresponding to the barrier which has been identified at low temperatures[93–95] (Sect. E.II.1., Fig. 26). A quantitative agreement of the calculated with the experimentally determined barrier heights, however, seems to be out of the scope of the discussed model.

The anisotropic continuum model allows to describe the occurrence of two distinctly different types of excited states, the free excitons and the strongly localized excitons. But this model cannot be expected to give a detailed energy level diagram. Assuming the D_{4h} symmetry group, the representations of the FE and STE states can be constructed from the representations of the relevant one-electron states[111]. For both FE and STE one obtains the representations $^1A_{2u}$ and $^3A_{2u}$ (ground state $^1A_{1g}$). The singlet-triplet splitting is expected to be larger for the STE than for the FE. Taking into account spin-orbit coupling one gets a mixing of the corresponding wavefunctions and a splitting of the states for both FE and STE into two sets of A'_{2u}, E'_u, and A'_{1u} (in D'_{4h}).

D.III. *The Model of Coupled $[Pt(CN)_4]^{2-}$ Ions*

Another attempt to describe the lowest excited states of the $[Pt(CN)_4]^{2-}$ chains starts with the many electron states of the isolated complex ions and investigates their mutual perturbations due to their coupling during the process of approach[85, 94, 117–119]. This method has the advantage that the excited states of the isolated complex ions and their energies are essentially known and that the intra complex electron-electron interaction in its main contribution has already been regarded. Due to the intrachain coupling the systems of degenerate excited states of the complex ions split into bands. With regard to the term system of the isolated $[Pt(CN)_4]^{2-}$ complex ion (see Table 3) three electronic

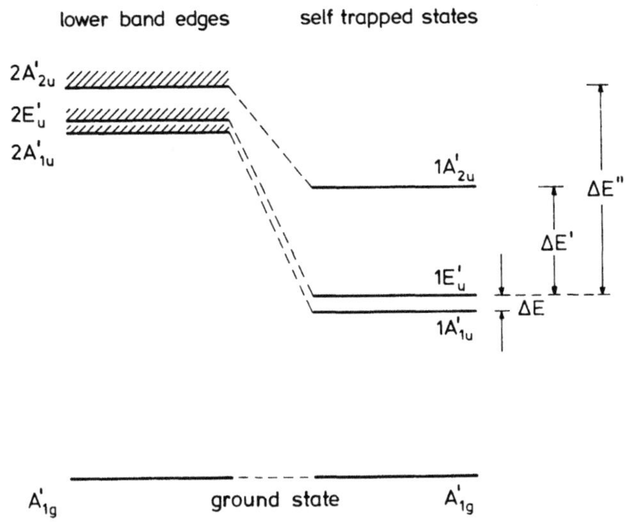

Fig. 13. Band structure and self-trapped states of a $[Pt(CN)_4]^{2-}$ chain (schematic, energies not to scale). The energetic order of A'_{1u} and E'_u is deduced from experimental results[94]. For a distinction of the free exciton states and the self-trapped states the symbols of the irreducible representations have been marked by additionally prefixed numbers 2 or 1, respectively

bands with the parentages A'_{1u}, E'_u, and A'_{2u}, respectively, are expected above the non-degenerate crystal ground state (A'_{1g}), as shown schematically in Fig. 13. In the case of self-trapping the states of the involved $[Pt(CN)_4]^{2-}$ ion (or of the small cluster of $[Pt(CN)_4]^{2-}$ ions) will be stabilized below the lower edges of the corresponding bands. Since the parentage states A'_{1u} and E'_u are derived from the same electron configuration and in particular from the same term $^3A_{2u}$, corresponding A'_{1u} and E'_u band states, especially the lower edges of the bands, and also the states of the self-trapped centers are anticipated to have nearly the same energies, respectively. Because of its different parentage ($^1A_{2u}$) the A'_{2u} gets a somewhat different energetic position. Since in the isolated complex ion the A'_{2u} state has a distinctly higher energy than the E'_u state ($\Delta \sim 2000\ cm^{-1}$, see Ref. 62) it may be assumed that a corresponding energetic order holds also for the lower band edges of the crystal states and also for the localized states of the self-trapped centers: $E(A'_{2u}) > E(E'_u), E(A'_{1u})$. The derived energy level diagram[94] has been proved to be effective for the interpretation of several optical properties, as shown in the following sections.

E. Polarized Spectroscopy at Different Temperatures

The optical spectroscopy between the near UV and the near IR represents an appropriate method to gain insight into the electronic properties of the crystals of the tetracyanoplatinates(II). Since the spectroscopic properties, such as oscillator strengths, transition energies, emission intensities show a strongly anisotropic behavior, the *polarized* spectroscopy yields the most comprehensive information.

E.I. Spectroscopic Properties Under Ambient Conditions

The various MCP compounds crystallize with different intrachain metal-metal distances R and exhibit drastic differences in their colors (see Table 1 and 5). The correlation between transition energies and R has already been expressed in 1951 by Yamada[141)] and later on by Moreau-Colin[13)]. Before this R-dependence is discussed in detail, typical spectra of a tetracyanoplatinate crystal will be presented.

Figure 14 reproduces the polarized reflectance spectra of a BaCP single crystal and the corresponding absorption spectra which have been calculated from the reflectance spectra by a Kramers-Kronig transformation[135, 136, 138)]. The maximum of the reflectivity reaches about 75% at 23,400 cm^{-1}. This high reflectivity value corresponds to an oscillator strength of the order of one (peak value of the molar extinction coefficient $\approx$ 80,000 l/mol · cm), indicating that the transition is strongly allowed. The energy of the calculated absorption maximum ($\approx$ 22,600 cm^{-1}) is slightly red-shifted. Kramers-Kronig transformations for other MCP compounds are not available. However, the energy difference between the measured reflectance and the calculated absorption maxima will be in the same order of magnitude for all MCP compounds. In the following for several comparisons the transition energies of the dominant $\vec{E} \parallel \vec{c}$ polarized peaks will be approximated by the energies of the reflectance maxima (see also Ref. 129). In the high energy region the spectra show some structure and are mainly perpendicularly polarized. The corresponding transitions are shortly discussed in Sect. E.I.2.

In Fig. 15 the measured low energy absorption and reflectance spectra are compared with the emission spectra of BaCP. At room temperature four bands can be resolved.

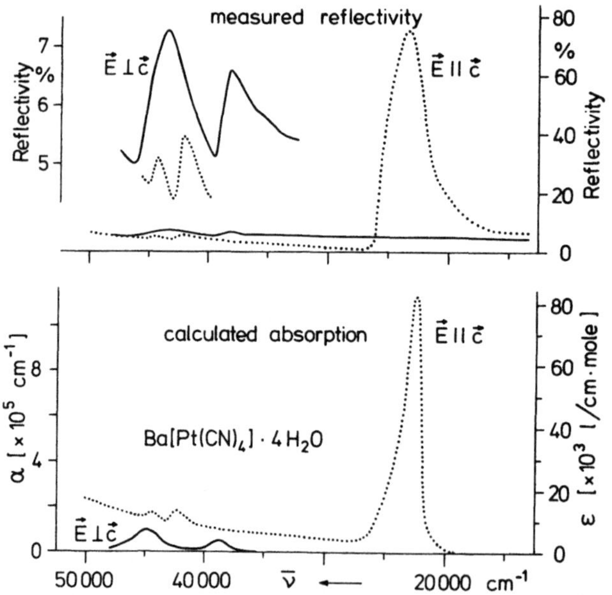

Fig. 14. Polarized reflectance and absorption spectra of single-crystal $Ba[Pt(CN)_4] \cdot 4\ H_2O$ (295 K, 1 bar). The $\vec{E} \parallel \vec{c}$ polarized reflectance spectra and the corresponding calculated absorption spectra are taken from Ref. 136, the $\vec{E} \perp \vec{c}$ polarized ones from Ref. 135, 138. The high energy structures in the reflectance spectra are enlarged by a factor of 10 (use left hand scale)

The $\vec{E}\|\vec{c}$ polarized absorption has a very high oscillator strength. Thus, only the flanks can be recorded. For BaCP at room temperature one could not even register the high energetic flank. However, for most other MCP compounds and also for BaCP at low temperatures the high energetic flanks have been resolved (see the Figs. 17 and 27). Because of its high oscillator strength the $\vec{E}\|\vec{c}$ polarized absorption is assigned to transitions between the singlet ground state A'_{1g} and the free exciton state of symmetry 2 A'_{2u}, which is mainly of singlet character (see Sects. D.II. and D.III.).

The $\vec{E}\perp\vec{c}$ polarized absorption is relatively weak, its oscillator strength has the order of 10^{-3} [117]. The $\vec{E}\perp\vec{c}$ peak has the same energy as the low energy flank of the strong $\vec{E}\|\vec{c}$ absorption (Fig. 15). This is valid, independent of the R values and of the temperature (see Figs. 17 and 27). The relatively small oscillator strength of the $\vec{E}\perp\vec{c}$ polarized transition and its energetically close connection to the low energy flank of the $\vec{E}\|\vec{c}$ polarized transition led to a classification of the corresponding excited state as a component of the free exciton state of symmetry E'_u being mainly of triplet character (see Sects. D.II. and D.III.).

The $\vec{E}\|\vec{c}$ polarized emission peak lies also in the flank of the $\vec{E}\|\vec{c}$ polarized absorption[1]. This emission is assigned to result from a singlet exciton state 2 A'_{2u}. The allowed-

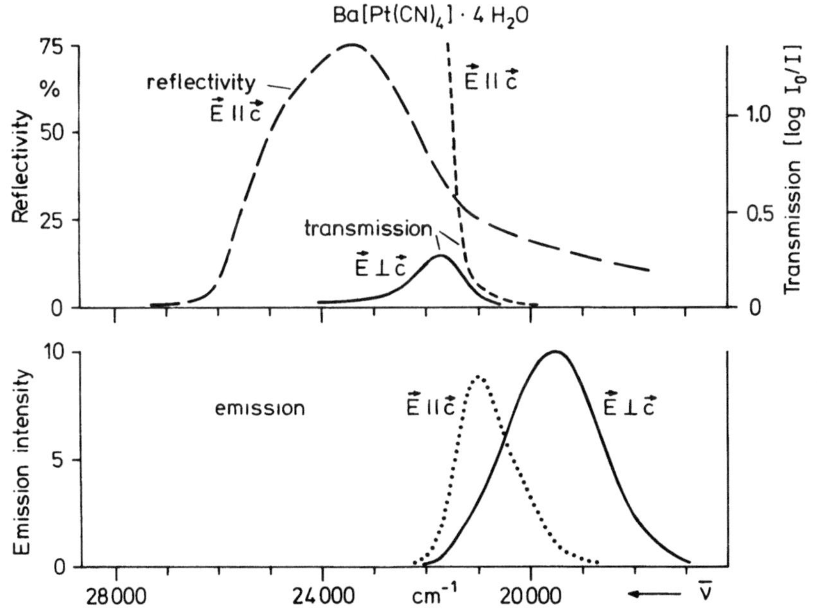

Fig. 15. Polarized spectra of single-crystal $Ba[Pt(CN)_4] \cdot 4\ H_2O$ (295 K, 1 bar)[30,85,113]. The intensities of the differently polarized emission spectra can be compared. The spectra are not corrected, neither due to the response of the apparatus nor due to reabsorption effects. Excitation: λ = 364 mm, $\vec{E}\|\vec{c}$; spectral resolution: 10 cm^{-1}; crystal thickness for the absorption measurements 10 μm

1 The strong spectral overlap of the $\vec{E}\|\vec{c}$ polarized emission with the corresponding absorption leads to reabsorption effects which cause a red-shift of the peak maxima of the $\vec{E}\|\vec{c}$ polarized emission. This effect can be reduced by excitation with $\vec{E}\|\vec{c}$ polarized light.

Table 5. Transition energies for different single-crystal tetracyanoplatinates(II) at T = 295 K (p = 1 bar). Low-temperature data are summarized in Table 7

Compound	Emission $[cm^{-1}]$ $\vec{E}\parallel\vec{c}$	$\vec{E}\perp\vec{c}$	Ref.	Reflectance $[cm^{-1}]$ $\vec{E}\parallel\vec{c}$	Ref.	Absorption peak energies $[cm^{-1}]$ $\vec{E}\perp\vec{c}$[b]	$\vec{E}\parallel\vec{c}$[c]	$\vec{E}\parallel\vec{c}$	$\vec{E}\perp\vec{c}$[d]	Ref.
$Sr[Pt(CN)_4] \cdot 3\ H_2O$	15,600[e]		29, 122							
$Mg[Pt(CN)_4] \cdot 7\ H_2O$	17,550	16,800	121, 123–129	19,400	129	17,900	24,800		39,215 43,480	
$Y_2[Pt(CN)_4]_3 \cdot 21\ H_2O$	17,800	16,950	13, 41, 127, 130	19,700	130				38,910 44,640	
$Tb_2[Pt(CN)_4]_3 \cdot 21\ H_2O$[a]	17,800	16,800	30	19,700	30					
$KLi[Pt(CN)_4] \cdot 2\ H_2O$	18,500	17,500	29	21,050	68	19,400				
$Sr[Pt(CN)_4] \cdot 2\ H_2O$	19,500	18,550	120							
$KNa[Pt(CN)_4] \cdot 3\ H_2O$	19,450	18,400	13, 41, 122, 131	21,850	68	20,300				
$(NH_4)_2[Pt(CN)_4] \cdot 2\ H_2O$	19,800	18,750	13, 29, 41, 120							
$(enH_2)[Pt(CN)_4] \cdot 0.5\ H_2O$	19,800	18,400	37, 132			20,700				37
$LiRb[Pt(CN)_4] \cdot 2\ H_2O$	20,300	18,700	120							
$Ba[Pt(CN)_4] \cdot 4\ H_2O$	21,000	19,500	13, 41, 122, 124, 127, 129, 131, 133, 134	23,300	30, 129, 135–137	21,700			38,760 44,440	
								42,300 44,400	38,700 44,700	135, 136, 138
$K_2Sr[Pt(CN)_4]_2 \cdot 6\ H_2O$	21,000	19,450	29, 120							
$Sm_2[Pt(CN)_4]_3 \cdot 18\ H_2O$[a]	21,650	19,400	30							
$Mg[Pt(CN)_4] \cdot 4.5\ H_2O$	21,700	20,200	120, 125, 127							
$Ca[Pt(CN)_4] \cdot 5\ H_2O$	22,250	20,700	13, 41, 122–124, 127, 129	24,400	68, 129	22,700	31,000		38,460 43,480	
$Rb_2[Pt(CN)_4] \cdot 1.5\ H_2O$	23,000	21,100	13, 29, 120, 127, 131			24,400				
$K_2[Pt(CN)_4] \cdot 3\ H_2O$	23,700	21,900	41, 122, 127, 139							

$Cs_2[Pt(CN)_4] \cdot 1\,H_2O$	25,200	22,800	13, 29, 127, 131, 134, 140					
$Sr[Pt(CN)_4] \cdot 5\,H_2O$	25,500	23,000	13, 41, 122, 124, 127, 129, 131, 134	28,300	129	26,900	35,100 38,600	38,460 44,250
$Rb_2[Pt(CN)_4] \cdot 1\,H_2O$	26,700	23,700	120					38,170 44,840
$Na_2[Pt(CN)_4] \cdot 3\,H_2O$	26,750	24,400	29, 122			28,500	39,300	38,760 45,920
exp. error	± 100			± 300		± 150		

[a] Data for further rare earth compounds are found in Tables 1 and 11; [b] The peak lies in the low-energetic flank of the highly allowed $\vec{E}\|\vec{c}$ polarized transition. See for example Fig. 15, Ref. 85, 120 and for MgCP[117, 121]; [c] The transition energy exhibits nearly the same R-dependence as the highly allowed $\vec{E}\|\vec{c}$ polarized one. See for example Fig. 17; [d] Polarization not clearly defined, only data of intense transitions are reproduced (taken from Ref. 13 unless stated otherwise); [e] unpolarized

ness of this transition is also displayed by the very short emission lifetime ($\tau \lesssim 300$ ps[142]), see also Sect. E.IV.).

The $\vec{E}\perp\vec{c}$ polarized emission exhibits a very different behavior. For example, the Stokes shift between the $\vec{E}\perp\vec{c}$ polarized absorption and emission is relatively large and R-dependent. Further, the $\vec{E}\perp\vec{c}$ emission liftetime is some orders of magnitude longer than for the high energetic $\vec{E}\|\vec{c}$ polarized emission. Therefore, it is expected that this emission results from a state mainly of triplet character. (At low temperature one can resolve further components in this energy region, see Sect. E.II.)

E.I.1. Low Energy Transitions at Different Pt–Pt Distances

A deeper understanding of the spectroscopic behavior is achieved when the Pt–Pt distance R is varied. Figure 16 shows the emission and reflectance spectra and Fig. 17 the absorption spectra for different MCP compounds. The corresponding peak energies are summarized in Table 5.

Red Shift

The main effect of a R reduction is the drastic red shift of the low energy transitions. Figure 18 shows the differently polarized emission, the $\vec{E}\perp\vec{c}$ polarized absorption and the $\vec{E}\|\vec{c}$ polarized reflectance energies of the peak maxima versus R^{-3}. From these plots one finds the following empirical equations:[29, 113, 122]

emission: $$\bar{\nu}_e(\vec{E}\|\vec{c}) = 42.9 \cdot 10^3 - 8.0 \cdot 10^5\, R^{-3} \quad (7)$$

$$\bar{\nu}_e(\vec{E}\perp\vec{c}) = 36.8 \cdot 10^3 - 6.3 \cdot 10^5\, R^{-3} \quad (8)$$

reflectance: $$\bar{\nu}_r(\vec{E}\|\vec{c}) = 45.5 \cdot 10^3 - 8.0 \cdot 10^5\, R^{-3} \quad (9)$$

absorption: $$\bar{\nu}_a(\vec{E}\perp\vec{c}) = 43.9 \cdot 10^3 - 8.0 \cdot 10^5\, R^{-3} \quad (10)$$

$\bar{\nu}$ is measured in cm^{-1} and R in Å. The experimental error of the constants is less than $\pm 0.5\%$ for Eqs. (7) and (8) and less than $\pm 1\%$ for Eqs. (9) and (10).

Especially, Eq. (7) is very useful. As shown in the Sects. E.II. and F.I. it can be applied approximately to determine R as well at low temperature as under high pressure just from measuring the emission spectra. Utilizing Eq. (7) also a number of structural informations has been obtained (see Sect. B and the Refs. 29, 30, 113, 130, 140, 143, 144).

In order to find a functional relation between R and the transition energies, attempts were made to apply the Frenkel exciton theory[145–148] to a single chain[117–119, 121]. From this theory, using a dipole-dipole approximation, *formally* a R^{-3} dependence results[119, 122, 149, 150]. From Fig. 18 it is seen that the plot of the transition energies versus R^{-3} actually yields a very useful *empirical* relation between R and the polarized transition energies[29, 68, 113, 122] (see Sect. D.).

Within limits of experimental error the $\vec{E}\|\vec{c}$ polarized reflectance and emission have the same R dependence as the relatively weak $\vec{E}\perp\vec{c}$ polarized *absorption*. (For all MCP's this peak lies in the flank of the $\vec{E}\|\vec{c}$ polarized, highly allowed absorption.) The R dependence of the $\vec{E}\perp\vec{c}$ polarized *emission* energy is substantially weaker. This behavior

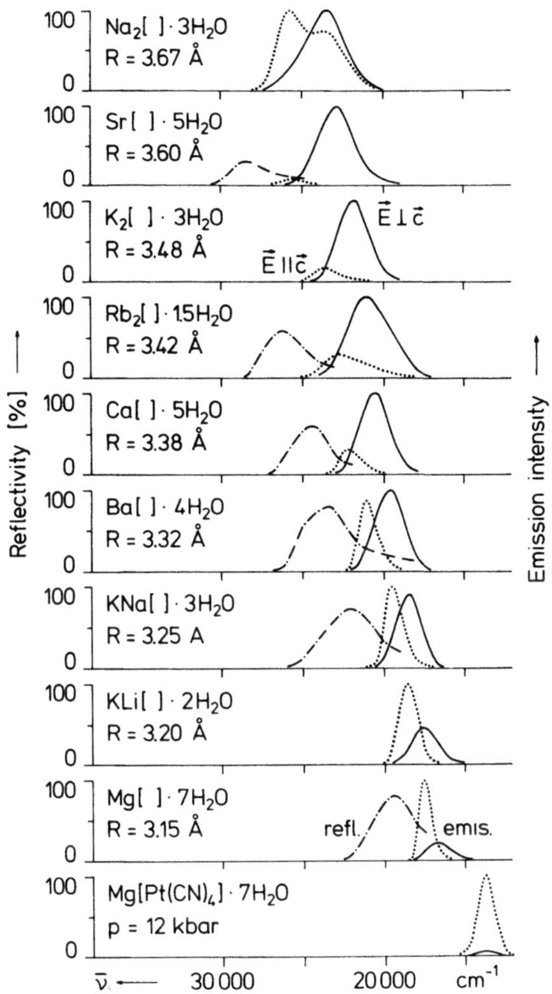

Fig. 16. Polarized reflectance ($\vec{E}\|\vec{c}$) and emission ($\vec{E}\|\vec{c}$, $\vec{E}\perp\vec{c}$) spectra of different single-crystal tetracyanoplatinates(II) at 295 K. For comparison the emission spectra of MgCP at 12 kbar are also reproduced. The emission intensities for different salts (and pressures) are not comparable. The excitation wavelengths (polarization: $\vec{E}\|\vec{c}$) correspond approximately to the maxima of reflectivity. The absolute reflectivity error is $\pm 15\%$. References are given in Table 5

points to the fact that the corresponding electronic states are of different origin. As discussed in Sects. E.II. and D the $\vec{E}\perp\vec{c}$ emission results from self-trapped states being mainly of triplet character. The R-value corresponding to the formal crossing of the differently polarized emissions is determined to ≈ 3.03 Å, cf. Eqs. (7) and (8). Approximately at this Pt–Pt distance, presumably, the self-trapping is no longer possible because the chains have become too stiff (see Sects. D.II. and F.I.).

Quasi One-Dimensionality

The above mentioned relations between the transition energies and the R values are valid for the different MCP compounds, although these compounds have – besides their com-

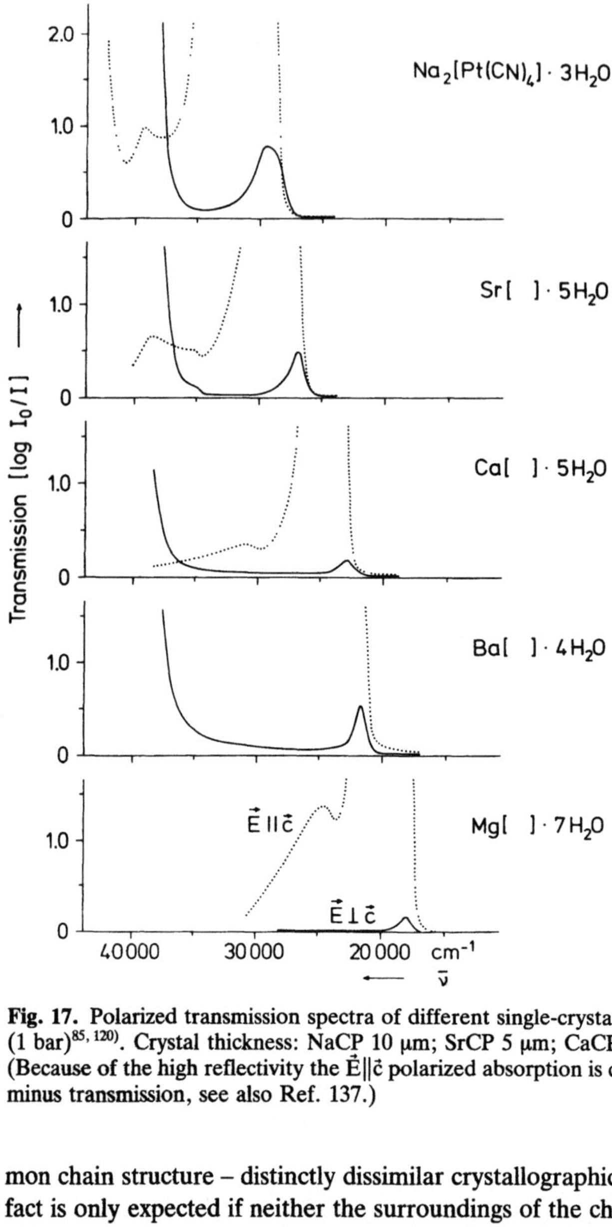

Fig. 17. Polarized transmission spectra of different single-crystal tetracyanoplatinates(II) at 295 K (1 bar)[85, 120]. Crystal thickness: NaCP 10 μm; SrCP 5 μm; CaCP 1 μm; BaCP 5 μm; MgCP 7 μm. (Because of the high reflectivity the $\vec{E} \| \vec{c}$ polarized absorption is calculated as one minus reflectivity minus transmission, see also Ref. 137.)

mon chain structure – distinctly dissimilar crystallographic structures (see Sect. B). This fact is only expected if neither the surroundings of the chains nor the exact structure of the chain itself – except the Pt–Pt distance R – strongly influence the corresponding transition energies. In this sense one may speak of quasi one-dimensional compounds.

E.I.2. Higher Energy Transitions

Besides the transitions discussed above further absorptions have been observed at higher energies (see Figs. 14, 17, 27, and Table 5). Transitions are found in both polarizations, however, at different energies and with dissimilar dependences on the Pt–Pt distance.

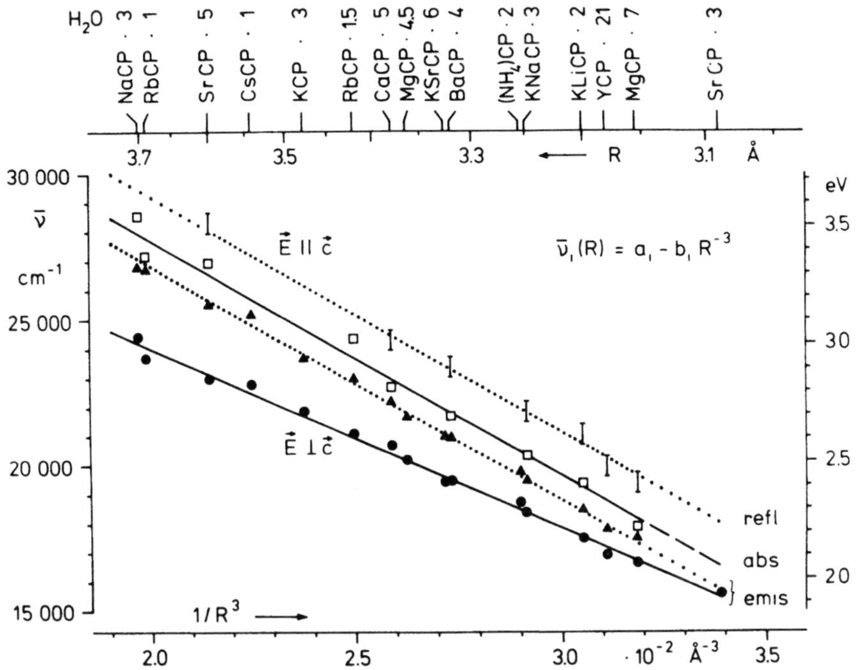

Fig. 18. Peak maxima of the polarized emission ($\vec{E}\|\vec{c}$, $\vec{E}\perp\vec{c}$), absorption ($\vec{E}\perp\vec{c}$) and reflectance ($\vec{E}\|\vec{c}$) versus $1/R^3$ for different $M_x[Pt(CN)_4] \cdot n\ H_2O$ single crystals (295 K, 1 bar)[29, 113, 122]. R is the Pt–Pt distance in the chain direction. Only the low-energetic transitions are used for this diagram. Experimental error: emission ± 100 cm^{-1}; absorption ± 150 cm^{-1}; reflectance ± 300 cm^{-1}. See also Eqs. (7) to (10)

$\vec{E}\perp\vec{c}$ Polarized Absorptions

Above 37,000 cm^{-1} there are several relatively strong $\vec{E}\perp\vec{c}$ polarized absorption peaks. Figure 17 shows only their absorption flanks but the Kramers-Kronig transformation of the reflectance spectra, which has been determined[135] for BaCP, displays two peaks lying about 6000 cm^{-1} apart (Fig. 14). These transitions do not exhibit a distinct R dependence (see also Ref. 13). Because of their energetic positions, their relatively high oscillator strengths, and their polarization properties these transitions have been assigned as spin allowed $e_g \rightarrow a_{2u}$ transitions within a single ion[13, 135]. Since the $e_g(d_{xz}, d_{yz})$ orbitals exhibit a substantially smaller inter complex overlap than the $a_{1g}(d_{z^2})$ orbitals the solid state interactions will be distinctly less pronounced.

The relatively weak $\vec{E}\perp\vec{c}$ polarized transitions near 35,000 cm^{-1} which are observed as shoulders at low temperature (Fig. 27) may result from parity or spin forbidden transitions. A series of such transitions is expected already for the single complex ion (see Table 3).

$\vec{E}\|\vec{c}$ Polarized Absorptions

All MCP compounds, which have been investigated by polarized absorption measurements, show a $\vec{E}\|\vec{c}$ polarized high energy transition with an oscillator strength of about three orders of magnitude smaller than that of the dominant peak[85, 120]. This transition occurs by about 6000 cm^{-1} higher than the center of the dominant $\vec{E}\|\vec{c}$ polarized peak. Both transitions exhibit nearly the same R dependences (see Fig. 17 and Table 5). The origin of this high energy transition is not yet fully understood.

E.II. Spectroscopic Properties at Low Temperatures

The spectroscopic properties of the MCP compounds vary strongly with decreasing temperature[2]. One finds a very large shift of the transition energies, a reduction of the half-widths, a change of the emission intensities and lifetimes, and for some compounds structural transformations.

E.II.1. Polarized Emission

The factor group which describes the symmetry of the solid state is different for nearly every MCP compound (see also Sect. B). For the understanding of the main optical properties the symmetry of the $[Pt(CN)_4]^{2-}$ chains can be approximated by the D'_{4h} group. Further, it is assumed that the symmetry group of the excited system, including the relaxed situation, is also D'_{4h}[37, 93, 94, 111, 113, 122, 129, 151].

As the ground state – which is of mainly singlet character – has the symmetry representation A'_{1g} (see Sects. C and D and Ref. 152) the selection rules for electric dipole transitions within the D'_{4h} symmetry are as follows: $E'_u \leftrightarrow A'_{1g}$ allowed for $\vec{E}\perp\vec{c}$, $A'_{2u} \leftrightarrow A'_{1g}$ allowed for $\vec{E}\|\vec{c}$.

The discussion of the experimental results can be based on the energy level diagram presented in Fig. 13.

Temperature reduction from ≈ 10 K to 1.9 K results in pronounced changes of the emission properties of the MCP compounds. These changes are similar for all compounds. Therefore, we restrict our discussion to two typical examples: BaCP and KLiCP. We start, reviewing the properties of the low energy emission with main emphasis given to the $\vec{E}\perp\vec{c}$ polarized data. (The low energy $\vec{E}\|\vec{c}$ polarized emission is often relatively weak compared to the $\vec{E}\perp\vec{c}$ polarized one and the available information is incomplete. Further, depolarization effects may introduce experimental artefacts.) Cooling from $T \approx 10$ K to 1.9 K exhibits the following effects[37, 93–95, 151]:

- The $\vec{E}\perp\vec{c}$ polarized emission component undergoes a red-shift of ≈ 300 cm^{-1} for BaCP and ≈ 200 cm^{-1} for KLiCP (see Figs. 19, 20, 24, and 25).
- The total emission intensity is reduced by a factor of four to five (Fig. 21).
- The emission decay time increases by about three orders of magnitude (Fig. 22 and Sect. E.IV.).
- At low temperatures (e.g. T = 1.9 K) magnetic fields induce a blue shift by 200 to 300 cm^{-1} of the low energetic components and further the emission decay time is reduced by about three orders of magnitude. At temperatures above about 10 K the emission properties do not depend on magnetic fields[94, 151]. (These effects are further discussed in Sect. G.)

---

2 Measurements on higher than ambient temperatures are not reported in the literature. At these temperatures the crystals partly lose their crystal water.

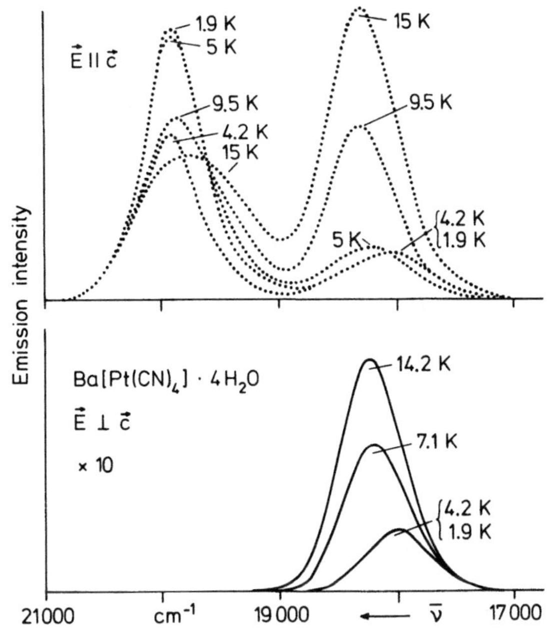

Fig. 19. Polarized emission of single-crystal $Ba[Pt(CN)_4] \cdot 4\ H_2O$ at different temperatures[94, 151]. The intensities for the different polarizations are comparable, if the $\vec{E} \perp \vec{c}$ polarized spectra are multiplied by a factor of 10. Excitation: 364 nm ($\vec{E} \| \vec{c}$)

From all these experimental results it follows that the lowest excited electron states of the self-traps are mainly of triplet character and have the symmetries A'_{1u} and E'_u, respectively[37, 93–95, 151, 30]. (See also Fig. 13.) The transition between $1A'_{1u}$ and A'_{1g} is strongly forbidden. Due to vibronic coupling, however, relatively weak transitions from the $1A'_{1u}$

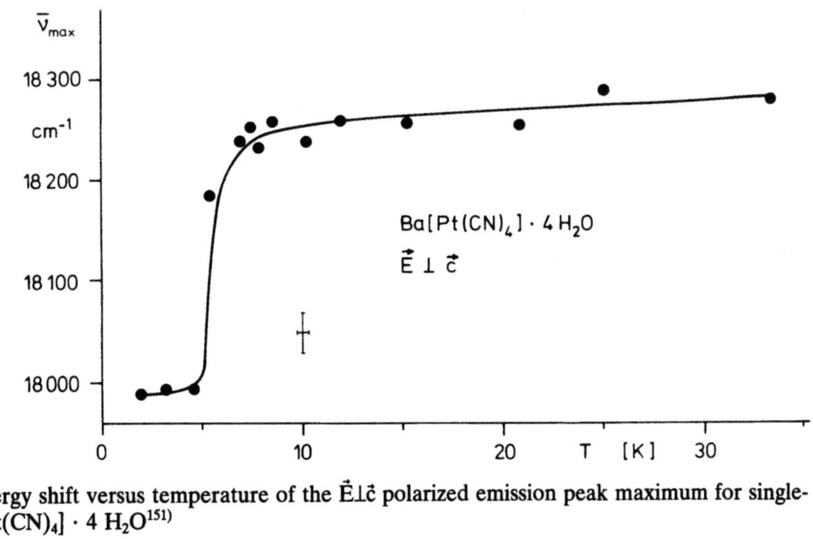

Fig. 20. Energy shift versus temperature of the $\vec{E} \perp \vec{c}$ polarized emission peak maximum for single-crystal $Ba[Pt(CN)_4] \cdot 4\ H_2O$[151]

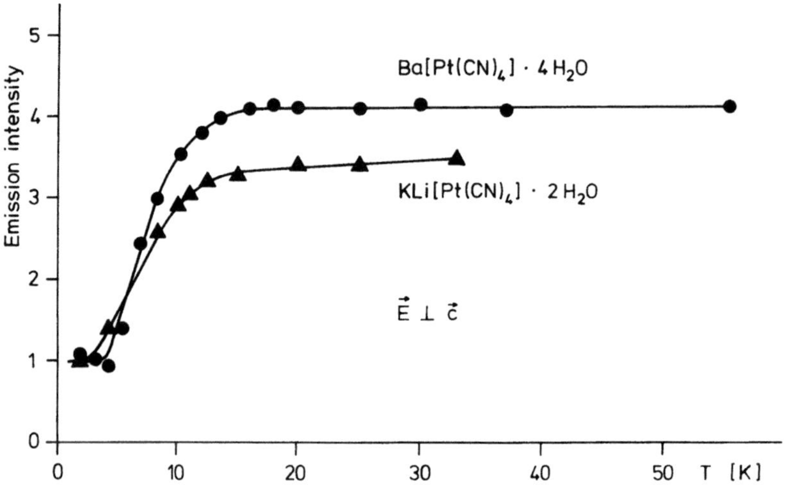

Fig. 21. Intensities of the $\vec{E} \perp \vec{c}$ polarized emission components of single-crystal $KLi[Pt(CN)_4] \cdot 2\,H_2O$ and $Ba[Pt(CN)_4] \cdot 4\,H_2O$ versus temperature[151]. The intensities at T = 1.9 K are normalized to one. Excitation wavelength: KLiCP 457.9 nm; BaCP 364 nm ($\vec{E} \| \vec{c}$)

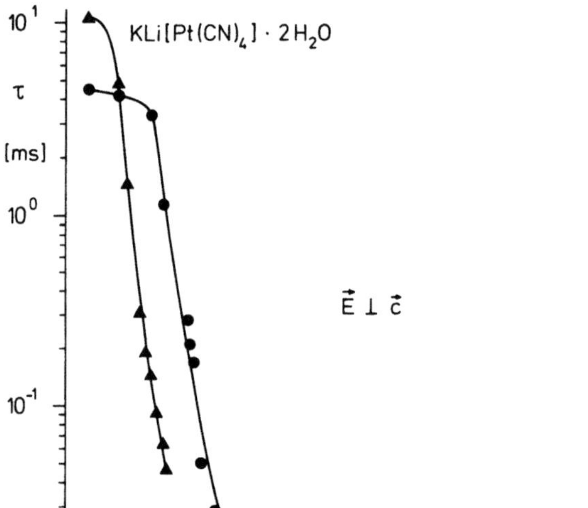

Fig. 22. Emission lifetime versus temperature for $KLi[Pt(CN)_4] \cdot 2\,H_2O$ and $Ba[Pt(CN)_4] \cdot 4\,H_2O$[94, 151]. The decay time is recorded only for the $\vec{E} \perp \vec{c}$ polarized emission component

state to excited vibrational levels of the A'_{1g} ground state become allowed, namely a $\vec{E}\perp\vec{c}$ transition with a ε_g vibration and a $\vec{E}\|\vec{c}$ transition with a α_{2g} vibration.

At very low temperature the transitions $1A'_{1u} \rightarrow A'_{1g} + \varepsilon_g(\vec{E}\perp\vec{c})$ and $1A'_{1u} \rightarrow A'_{1g} + \alpha_{2g}(\vec{E}\|\vec{c})$ are expected. With increasing temperature, however, the $1E'_u$ state is populated thermally. From this state the $\vec{E}\perp\vec{c}$ emission into the vibrationally non-excited ground state A'_{1g} is allowed. This explains the above mentioned temperature dependence of the emission energy, the intensity, and the emission decay time. The interpretation of the magnetic field effects follows from analogous arguments, as discussed in Sect. G. With the wavenumber of the $[Pt(CN)_4]^{2-}$ ε_g-vibration (321 cm^{-1}, Ref. 153) and a $1A'_{1u} - 1E'_u$ energy distance of some tens of wavenumbers (MgCP, KLiCP: 30 ± 5 cm^{-1}; BaCP, CaCP: 50 ± 5 cm^{-1}) good quantitative agreement with the experimental data is found[94].

The maximum of the very weak, low energetic $\vec{E}\|\vec{c}$ polarized emission, probably due to the transition $1\,A'_{1u} \rightarrow A'_{1g} + \alpha_{2g}$, is found at 1.9 K for BaCP as well as for KLiCP at nearly the same energy as the corresponding $\vec{E}\perp\vec{c}$ polarized component[93, 94]. With temperature increase to $T \approx 10$ K both components shift to higher energies (as described above) but with different values. The shift of the $\vec{E}\|\vec{c}$ polarized maximum is by about 100 cm^{-1} more pronounced (see Fig. 19). This larger blue shift is not yet fully understood. Possibly, it is brought about by an overlapping emission resulting from the higher lying $1A'_{2u}$ state, which is considered below.

Temperature increase above ≈6 K leads to a further emission band which is only observed in the parallel polarization (see the intensity growing at the low energy flank of the $\vec{E}\|\vec{c}$ emission peak near 20,000 cm^{-1} in Fig. 19). Therefore, the corresponding state is assigned to A'_{2u} symmetry. Time-resolved emission spectroscopy[154], by which the transition could clearly be separated from the other emission components, has shown that its decay time is less than 3 ns, indicating a high amount of singlet character. From the theoretical investigations, presented in Sect. D, it seems possible that the emitting A'_{2u} corresponds to the self-trapped exciton state $1A'_{2u}$ of predominant singlet character, as shown in Fig. 13. Corresponding peaks have been observed for KNaCP (at $\Delta E' \approx$ 500 cm^{-1} [68]), for BaCP (≈ 1000 cm^{-1} [68, 151]), and for CaCP (≈ 1600 cm^{-1} [158]). For BaCP the relative intensity of this transition strongly depends on the wavelength of excitation (see Fig. 23), showing that different processes govern the relaxation cascade down from the absorbing states[68].

The emission of highest energy is always $\vec{E}\|\vec{c}$ polarized and therefore is assigned to emerge from the $2A'_{2u}$, cf. Fig. 13[113, 121, 122]. Its decay time is shorter than 0.3 ns[94, 142]. This small value coupled with the relatively high emission intensity indicates a high oscillator strength of this transition $A'_{1g} \leftrightarrow 2A'_{2u}$. The emission from $2A'_{2u}$ can be observed over the whole temperature range (see Figs. 24 and 25), exhibiting an unusual behavior of the emission decay properties and of the temperature dependence of the emission intensity.

Figure 26 reproduces for KLiCP, as a typical example, the intensity ratio $I(\vec{E}\perp\vec{c})/I(\vec{E}\|\vec{c})$. It increases from ≈5 K to ≈75 K by a factor of 10 and then it decreases again from ≈75 K to 300 K by a factor of about 35[120, 120a]. This effect is rationalized by the assumption of an energy barrier between the $2A'_{2u}$ state and the very lowest excited states (see also Fig. 11)[68, 111, 155]. From Fig. 26 an activation energy of about 30 cm^{-1} is estimated for KLiCP. The occurrence of such a barrier can be explained also within a configuration coordinate diagram as a result of the self-trapping process[156]. With temperature increase from 1.9 K the barrier can be surmounted leading to a decrease of the

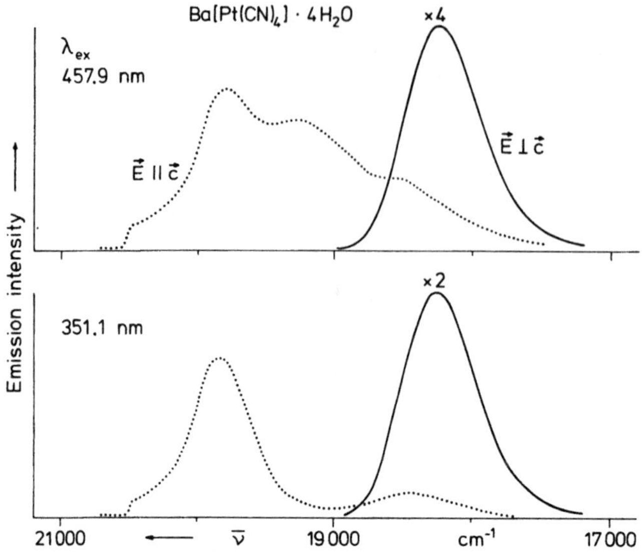

Fig. 23. Polarized emission of single-crystal $Ba[Pt(CN)_4] \cdot 4\ H_2O$ (at 4 K) for different excitation wavelengths (polarization of excitation: $\vec{E}\|\vec{c}$)[68]. The intensities of the spectra, excited with the same wavelength, are comparable

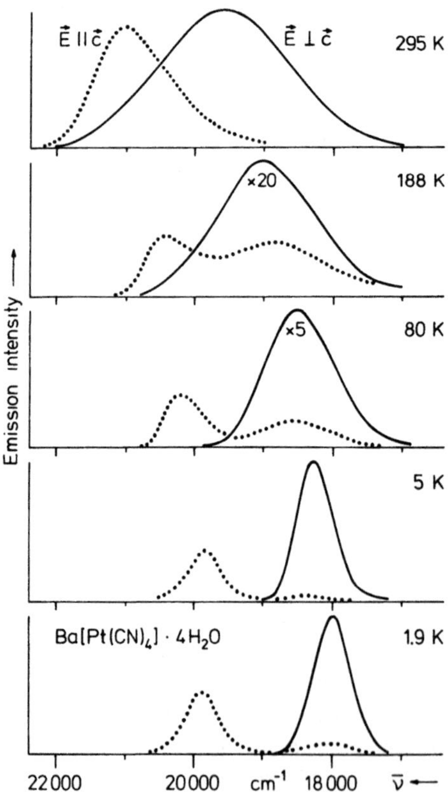

Fig. 24. Polarized emission of single-crystal $Ba[Pt(CN)_4] \cdot 4\ H_2O$ for different temperatures[68, 94]. The intensities of the spectra, recorded at the same temperature, are comparable, when multiplied by the given factors. Excitation: 364 nm ($\vec{E}\|\vec{c}$)

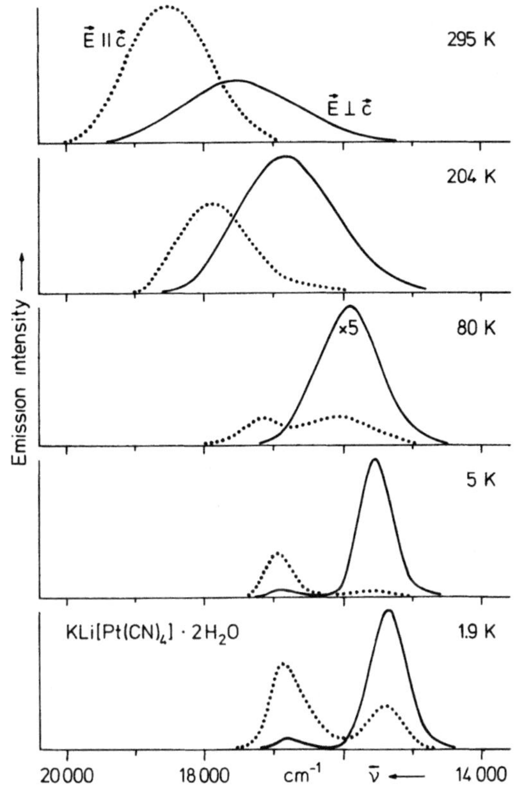

Fig. 25. Polarized emission of single-crystal KLi[Pt(CN)$_4$] · 2 H_2O for different temperatures[68, 120, 151]. The intensities of the spectra recorded at the same temperature are – besides the 1.9 K spectra – comparable. Excitation: 457.9 nm ($\vec{E}\|\vec{c}$)

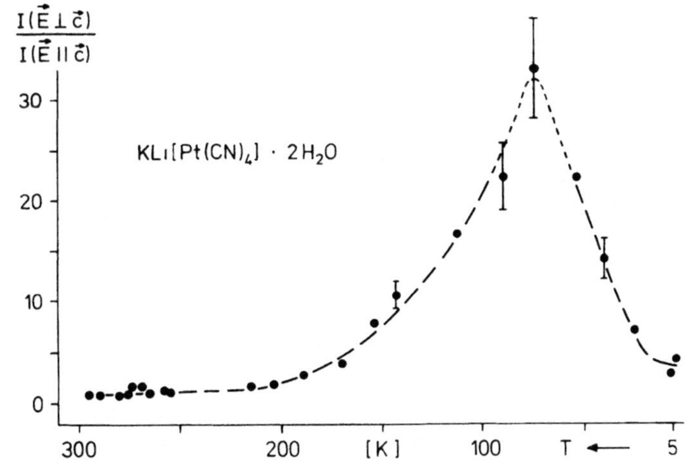

Fig. 26. Intensity ratio of the differently polarized emission of single-crystal KLi[Pt(CN)$_4$] · 2 H_2O versus temperature. I($\vec{E}\perp\vec{c}$) and I($\vec{E}\|\vec{c}$) belong to the low energy $\vec{E}\perp\vec{c}$ polarized and the high energy $\vec{E}\|\vec{c}$ polarized emission components, respectively. The plot results from data, taken from the Refs. 29, 120, 120a (see also Fig. 25)

Table 6. Experimentally determined activation energies for several $M_x[Pt(CN)_4] \cdot n\ H_2O$ crystals. ΔE and $\Delta E''$ are defined in Fig. 13

Compound	ΔE $[cm^{-1}]$	$\Delta E''$ $[cm^{-1}]$
$Mg[Pt(CN)_4] \cdot 7\ H_2O$	30[151]	620[68]
$Y_2[Pt(CN)_4]_3 \cdot 21\ H_2O$		890[68]
$KLi[Pt(CN)_4] \cdot 2\ H_2O$	30[151]	1050[68]
$Ba[Pt(CN)_4] \cdot 4\ H_2O$	52[151]	1470[68]
	94.8[157]	813[157]
$Ca[Pt(CN)_4] \cdot 5\ H_2O$	50[151]	1450[68]
$Na_2[Pt(CN)_4] \cdot 3\ H_2O$		2700[68]
experimental error	$\pm 5\ cm^{-1}$	$\pm 15\%$

intensity resulting from $2A'_{2u}$. In competition to this activated relaxation process one finds a spontaneous emission from $2A'_{2u}$, which can be registered in the whole temperature range.

With further temperature increase the relative $\vec{E}\|\vec{c}$ polarized intensity (from $2A'_{2u}$) goes up again, which can be explained by a thermal repopulation of $2A'_{2u}$. Consequently, the emission from $2A'_{2u}$ consists of two components, the short-living spontaneous one and the long-living repopulated component[68, 154], both exhibiting comparable intensities in the medium temperature range. However, above ≈ 250 K the repopulated and thus delayed emission predominates strongly. Therefore, one can determine the activation energy simply by an Arrhenius plot, using the temperature dependence of the intensity ratio[68, 113, 121]. From Fig. 26 it results for KLiCP an activation energy of 1050 cm^{-1} ($\pm$ 15%). Since in this temperature range (above ≈ 250 K) the $\vec{E}\perp\vec{c}$ polarized emission emerges mainly from $1E'_u$ the activation energy corresponds to $\Delta E''$ (Fig. 13). Further data are summarized in Table 6.

E.II.2. Polarized Absorption

Figure 27 reproduces the polarized absorption spectra of several MCP compounds, recorded at 10 K[85]. As expected, also at this temperature one cannot measure the transmission in the maximum of the dominant $\vec{E}\|\vec{c}$ polarized peak. This agrees with experimental results of Day and Ferguson[137], who observed an increase of the reflectivity of about 10–20% on temperature reduction from 295 K to 9 K. The low energy $\vec{E}\|\vec{c}$ absorption flank belongs to the transition between the A'_{1g} ground state and the exciton states at the lower edge of the $2A'_{2u}$ band, whereas the relatively weak $\vec{E}\perp\vec{c}$ peak corresponds to the spin forbidden transition $A'_{1g} \rightarrow 2E'_u$. Since NaCP is composed of two crystallographically non-equivalent chains several transitions occur in the corresponding energy range. The spectra of NaCP, SrCP, and CaCP show at the low energy flank of the $\vec{E}\|\vec{c}$ absorption bands an additional weak $\vec{E}\|\vec{c}$ absorption (for details see Ref. 85). Its energy coincides with the peak energy of the $\vec{E}\perp\vec{c}$ absorption. The appearance of this weak $\vec{E}\|\vec{c}$ absorption cannot be explained merely by geometrical depolarization, since there is no depolarization effect in the $\vec{E}\perp\vec{c}$ spectrum. Therefore, it may be ascribed to an additional $\vec{E}\|\vec{c}$ transition due to a lowering of the symmetry[85].

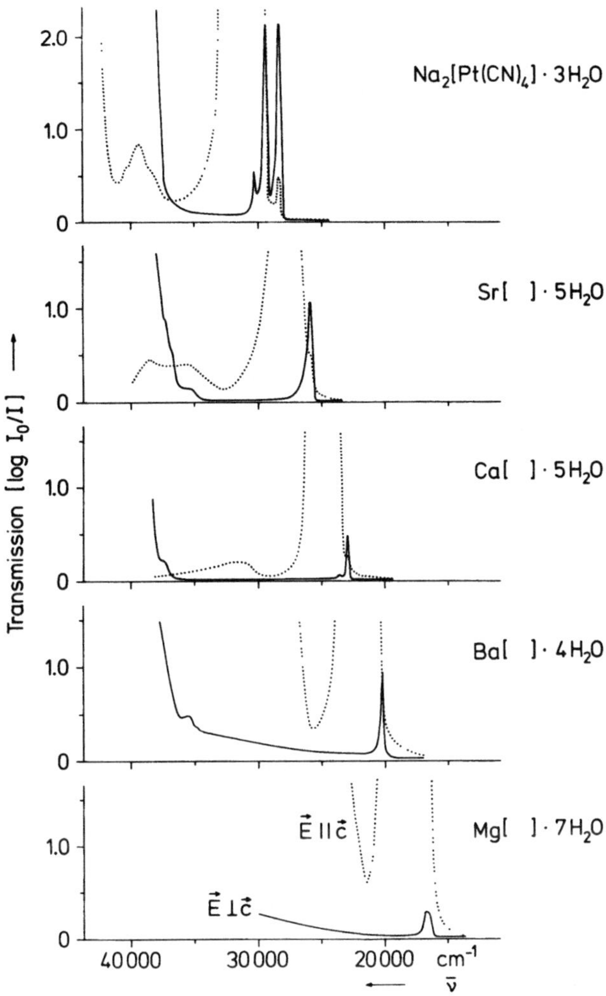

Fig. 27. Polarized transmission of different $M_x[Pt(CN)_4] \cdot n\ H_2O$ single crystals at 10 K[85]. The crystal thicknesses are given below Fig. 17

E.III. Red-Shift With Temperature Reduction

Figure 28 shows the temperature dependence of the emission energies of various MCP compounds with different Pt–Pt distances (see also the Figs. 20, 24, 25, and Table 7). In this section the spectral changes at temperatures below ≈ 10 K will not be regarded. These effects have a different physical origin and were discussed in Sect. E.II.1. Figure 28 also displays the temperature dependence of the $\vec{E} \perp \vec{c}$ polarized absorption for MgCP[121]. Its slope equals that of the high energy $\vec{E} \| \vec{c}$ polarized emission (resulting from $2A'_{2u}$, cf. Fig. 13), while the $\vec{E} \perp \vec{c}$ polarized emission (resulting mainly from $1E'_u$) does not correlate with the absorption.

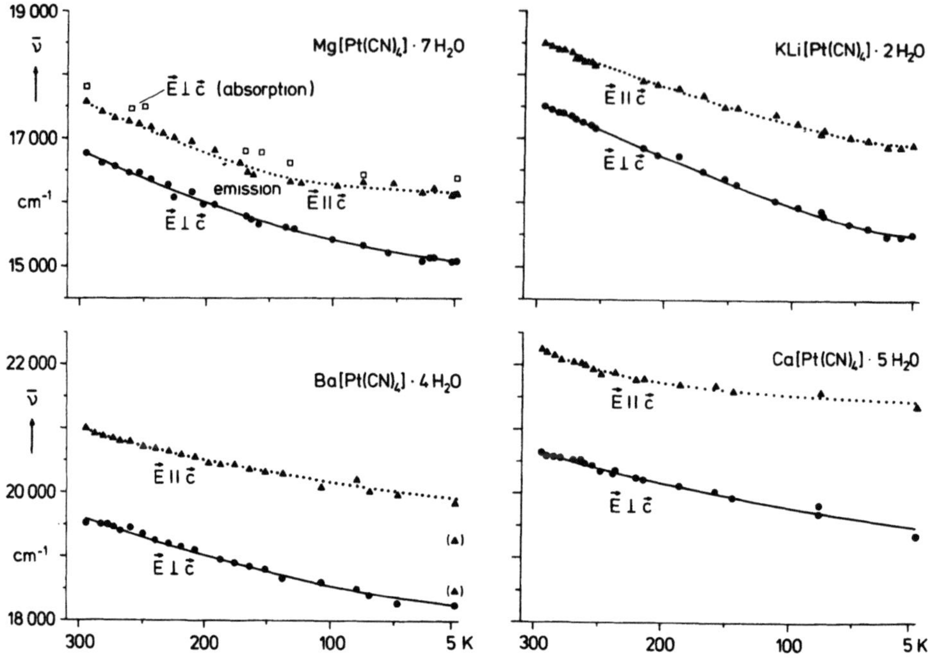

Fig. 28. Transition energies for different single-crystal tetracyanoplatinates(II) versus temperature[113, 68]. The plots represent the low energy $\vec{E}\perp\vec{c}$ polarized and the high energy $\vec{E}\|\vec{c}$ polarized emission peak maxima, respectively. For MgCP, also the temperature dependence of the $\vec{E}\perp\vec{c}$ polarized low-energy absorption peak is reproduced (taken from Ref. 121)

The red-shift of the transition energies under temperature reduction is explained in its main trend by a decrease of the Pt–Pt distance through thermal contraction leading to an enhancement of the intra-chain interaction[113, 121]. For KLiCP[160] and DyCP[43, 160] the temperature dependence of the Pt–Pt distance has been determined by X-ray diffraction and compared to the temperature dependence of R, which results from the emission peak energies according to the Eqs. (7) and (8), respectively. For the $\vec{E}\|\vec{c}$ polarized emission, good agreement is achieved; but using the perpendicularly polarized emission peak energies the calculated R values diverge increasingly the lower the temperature. The deviation becomes evident below $T \approx 200$ K[43, 68, 160]. The applicability of the Eqs. (9) and (10) at low temperature has not yet been investigated experimentally but it is reasonable to assume that they can also be used approximately.

From these investigations it follows that Eq. (7) may be utilized approximately to determine unknown low temperature R(T)-values and also the coefficients of thermal expansion ($\alpha_c\,[K^{-1}]$) in direction of the c-axis (see also Sect. F.I.):

$$\alpha_c = \frac{1}{R} \cdot \frac{\Delta R}{\Delta T} = \frac{R^3}{3 \cdot 8.0 \cdot 10^5} \cdot \frac{\Delta\bar{\nu}}{\Delta T} \tag{11}$$

$\Delta\bar{\nu}\,[cm^{-1}]$ is the value of red-shift of the $\vec{E}\|\vec{c}$ polarized emission peak maximum under temperature reduction by ΔT[K]. Table 8 summarizes α_c values, calculated by this equa-

Table 7. Transition energies for different single-crystal tetracyanoplatinates(II) at low temperatures (p = 1 bar). Room temperature data are summarized in Table 5

Compound	Emission [cm^{-1}] 80 K[a]		5 K[a]		1.9 K[b]		Absorption [cm^{-1}] 80 K[c]	10 K[d]		Ref.[e]
	$\vec{E}\parallel\vec{c}$	$\vec{E}\perp\vec{c}$	$\vec{E}\parallel\vec{c}$	$\vec{E}\perp\vec{c}$	$\vec{E}\parallel\vec{c}$	$\vec{E}\perp\vec{c}$	$\vec{E}\perp\vec{c}$	$\vec{E}\perp\vec{c}$	$\vec{E}\parallel\vec{c}$	
$Mg[Pt(CN)_4] \cdot 7\ H_2O$	16,350	15,400	16,100	15,100	16,150	14,850		16,400 16,700		121, 125, 128, 159
$KLi[Pt(CN)_4] \cdot 2\ H_2O$	17,150	15,900	16,950	15,550	16,850	15,350				
$KNa[Pt(CN)_4] \cdot 3\ H_2O$	18,500	16,950	18,150	16,800	18,150	16,550				
$(NH_4)_2[Pt(CN)_4] \cdot 2\ H_2O$	19,500	17,900								
$enH_2[Pt(CN)_4] \cdot 0.5\ H_2O$	18,850	17,400	18,900	17,200	19,000	17,100	19,650	19,550		37, 132
$LiRb[Pt(CN)_4] \cdot 2\ H_2O$	18,800	17,400								
$Ba[Pt(DCN)_4] \cdot 4\ H_2O$	20,200	18,450	19,900	18,300	19,800	18,000	20,200	21,700		30, 127, 137, 154, 155, 157, 159
$Sm_2[Pt(CN)_4]_3 \cdot 18\ H_2O$[f]	21,250	18,950								30, 143
$Mg[Pt(CN)_4] \cdot 4.5\ H_2O$	21,400	19,500								127
$Ca[Pt(CN)_4] \cdot 5\ H_2O$	21,600	19,850		19,300	21,400	19,300	21,700	22,900	31,500	127
$Rb_2[Pt(CN)_4] \cdot 1.5\ H_2O$	22,450	20,350					23,400			
$K_2[Pt(CN)_4] \cdot 3\ H_2O$	23,100	21,000								158
$Cs_2[Pt(CN)_4] \cdot 1\ H_2O$	25,100	22,350								140
$Sr[Pt(CN)_4] \cdot 5\ H_2O$	24,550	22,000						25,900	35,500 38,600	127
$Rb_2[Pt(CN)_4] \cdot 1\ H_2O$	26,300	22,600					27,100 28,200			
$Na_2[Pt(CN)_4] \cdot 3\ H_2O$	26,200	23,500				22,500		28,400 29,400 30,300	39,400	

[a] Ref. 29, 68, 113, 122; [b] Ref. 36, 94, 151; [c] Ref. 120; [d] The energetical red-shifts of the peak maxima with temperature reduction (compared to room temperature data, Table 5) may be reduced by the crystal supports, being of quartz glas with low thermal expansion. Data from the Refs. 85, 86; [e] These references refer to other publications which give low-temperature data; [f] Data for further rare earth compounds are found in Tables 1 and 11

Table 8. Coefficients of thermal expansion in the chain direction[68]

Compound	α_c[$\times 10^{-5}$ K^{-1}] (295 K)		
	Spectroscopic determination	Other Methods	Refs.
$Mg[Pt(CN)_4] \cdot 7\,H_2O$	14[a]		121
$Y_2[Pt(CN)_4]_3 \cdot 21\,H_2O$	12		113, 130
$Dy_2[Pt(CN)_4]_3 \cdot 21\,H_2O$	9	14 ± 2[b]	43, 160
$Er_2[Pt(CN)_4]_3 \cdot 21\,H_2O$	8	13.5 ± 2[b]	43, 160
$Tb_2[Pt(CN)_4]_3 \cdot 21\,H_2O$	8		160
$KLi[Pt(CN)_4] \cdot 2\,H_2O$	11	15 ± 2[b]	29, 160
$Ba[Pt(CN)_4] \cdot 4\,H_2O$	9[a]		
$Ca[Pt(CN)_4] \cdot 5\,H_2O$	11[a]		68
$K_2[Pt(CN)_4] \cdot 3\,H_2O$	–	11.8 ± 0.6[c]	161
$Na_2[Pt(CN)_4] \cdot 3\,H_2O$	8		68
$K_2[Pt(CN)_4] \cdot Br_{0.3} \cdot 3\,H_2O$	–	6.7 ± 0.2[b]	161, 162

[a] According to the right part of Eq. (11). The $\Delta\bar{\nu}/\Delta T$ values result for example from Fig. 28. Experimental error: ±30%; [b] From X-ray investigations; [c] From dilatometric investigations

tion for different MCP compounds and compares them with data found by other methods.

The Pt–Pt distance can be reduced by temperature reduction or by high pressure application (as comprehensively discussed in Sect. F). But there is one distinct dissimilarity concerning the energy difference between the two differently polarized main emission peaks. High pressure application *lowers* this energy difference (see Sect. F) while temperature reduction brings about an *increase* of the energy difference (see Fig. 28). Therefrom it follows that besides the Pt–Pt distance reduction the temperature dependence of the occupation of the electronic states is important.

E.IV. Kinetics of the Deactivation Processes

The energy level system shown in Fig. 29 proves to be an appropriate basis for the discussion of the kinetics of the optical emission processes. The treatment of the time behavior will be simplified, if the low temperature case and the high temperature case are discussed separately[93–95, 156, 157].

For $T \gtrsim 80$ K the temperature dependence of the emission is mainly determined by the thermal repopulation of the A'_{2u} states from the $1E'_u$ and the $1A'_{1u}$ states. For $kT \gg \Delta E$ the temperature behavior of the emission will not be influenced by the $1A'_{1u}$ state for the following reasons. The $1A'_{1u}$ deactivation rate into the ground state A'_{1g} is extremely small compared to the rate from $1E'_u$. Therefore, a direct emission from $1A'_{1u}$ is not expected and the deactivation of $1A'_{1u}$ consists mainly in a thermal repopulation of $1E'_u$. Hence, for the high temperature kinetics the $1A'_{1u}$ state can be neglected. With the transition channels described in Fig. 29 the simplified rate equations for the occupation numbers n_3 and n_2 of the states $2A'_{2u}$ and $1E'_u$, respectively, are[156]

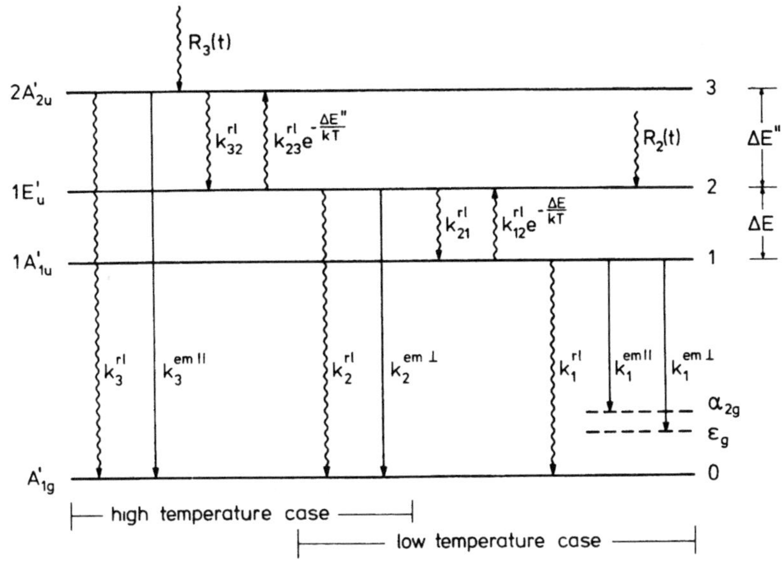

Fig. 29. Simplified energy level diagram serving as a basis for the discussion of the kinetics of the emission of tetracyanoplatinate(II) single crystals[94, 156]

$$\frac{dn_3}{dt} = -(k_3^{em} + k_3^{r\ell} + k_{32}^{r\ell})n_3 + k_{23}^{r\ell} \cdot e^{-\Delta E''/kT} \cdot n_2 + R_3(t) \tag{12}$$

$$\frac{dn_2}{dt} = -(k_2^{em} + k_2^{r\ell} + k_{23}^{r\ell} \cdot e^{-\Delta E''/kT}) \cdot n_2 + k_{32}^{r\ell} \cdot n_3 \tag{13}$$

$R_3(t)$ is the feeding rate of the level 3. The population of the $1E_u'$ level is reduced to a non-radiative process, and for the temperature dependence of the thermal repopulation of the $2A_{2u}'$ an exponential form is assumed[163].

For a description of the time behavior of the emission the $R_3(t)$ feeding rate is assumed to be an extremely short excitation pulse. After prompt relaxation into the $2A_{2u}'$ state the following time dependences of the occupation numbers result[156]:

$$n_3(t) \approx D_1 \cdot e^{K_- \cdot t} + \frac{k_{23}^{r\ell} \cdot e^{-\Delta E''/kT}}{k_3^{em} + k_3^{r\ell} + k_{32}^{r\ell}} D_2 \cdot e^{K_+ \cdot t} \quad (2A_{2u}' \text{ emission}) \tag{14}$$

$$n_2(t) \approx -\frac{k_{32}^{r\ell}}{k_3^{em} + k_3^{r\ell} + k_{32}^{r\ell}} D_1 \cdot e^{K_- \cdot t} + D_2 \cdot e^{K_+ \cdot t} \quad (1E_u' \text{ emission}) \tag{15}$$

with the parameters

$$K_+ = -\left[(k_2^{em} + k_2^{r\ell}) + (k_3^{em} + k_3^{r\ell}) \frac{k_{23}^{r\ell}}{k_{32}^{r\ell}} \cdot e^{-\Delta E''/kT}\right] \tag{16}$$

$$K_- = -\left[(k_3^{em} + k_3^{r\ell} + k_{32}^{r\ell}) + k_{23}^{r\ell}\left(1 - \frac{k_3^{em} + k_3^{r\ell}}{k_{32}^{r\ell}}\right) e^{-\Delta E''/kT}\right] \tag{17}$$

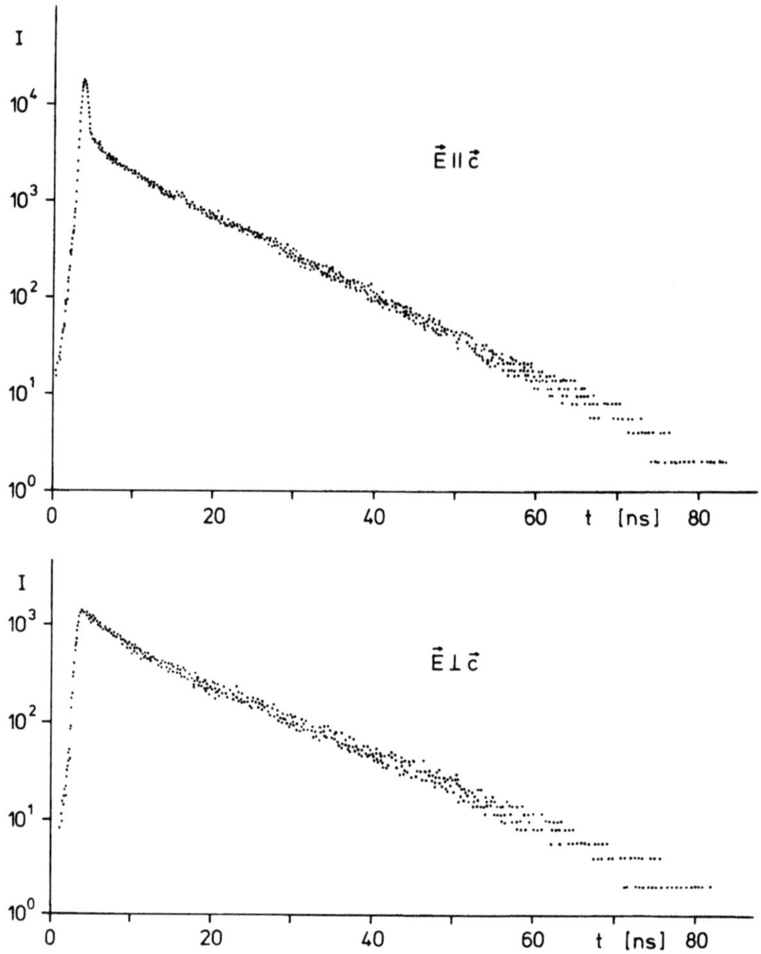

Fig. 30. Decay behavior of the $\vec{E}\|\vec{c}$ (at 17,600 cm^{-1}) and $\vec{E}\perp\vec{c}$ emission (at 16,900 cm^{-1}) of YbCP at T = 295 K. At this temperature the fast component is observed

$|K_+|$ is smaller than $|K_-|$, but the parameter K_+ exhibits a stronger temperature dependence than K_-. Both functions $n_3(t)$ and $n_2(t)$ are bi-exponential. $n_3(t)$ describes the decrease of the $2A'_{2u}$ emission, whereby the relative portion of the slow component increases with increasing temperature (see also Sect. E.II.1.). The emission from the $1E'_u$ state is composed of a fast increase, corresponding to K_-, and a slow decrease with the same time constant as the slow component of the $2A'_{2u}$ emission.

The experimental decay behavior of the $2A'_{2u}$ and the $1E'_u$ emission (Fig. 30) confirms the theoretical time dependence of the rates, given above. The fit curve for YbCP in Fig. 31 yields a good approximation of the experimental values.

For $T \lesssim 50$ K the thermal repopulation of $2A'_{2u}$ from $1E'_u$ and from $1A'_{1u}$ can be neglected because of the relatively large energy differences. Only the feeding rate $R_2(t)$ of $1E'_u$ will be regarded. Taking into account explicitly the small rate $R_1(t)$ of the $1A'_{1u}$, the results will be changed negligibly[156]. Thus the behavior of the occupations n_1 and n_2

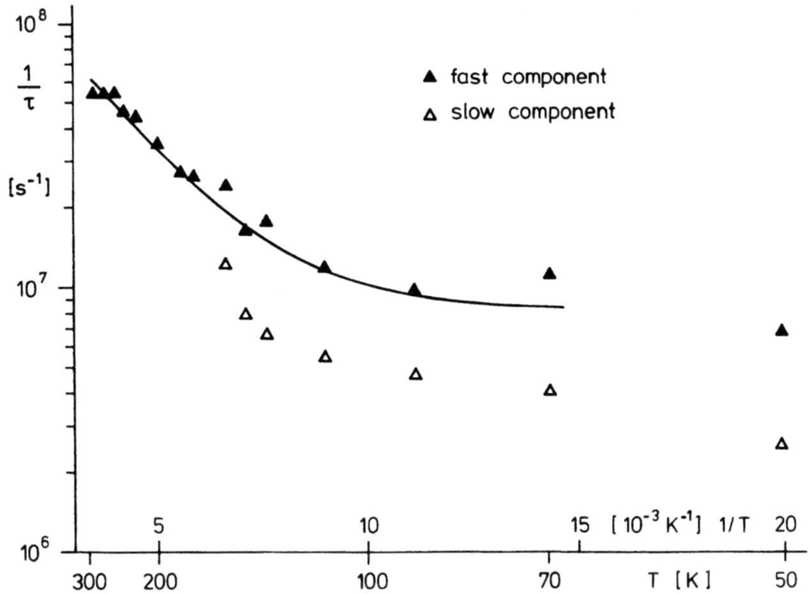

Fig. 31. Temperature dependence of the $\vec{E}\perp\vec{c}$ deactivation rate of YbCP in the high temperature range (T > 80 K). Fit curve for the fast component. Below T ≈ 150 K an additional slow component can be observed[156)]

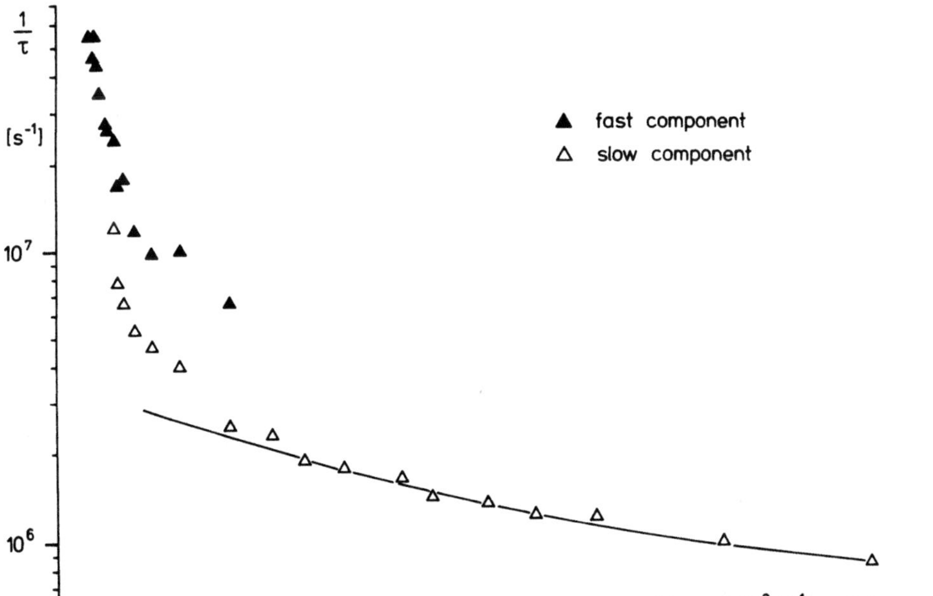

Fig. 32. Temperature dependence of the $\vec{E}\perp\vec{c}$ deactivation rate of YbCP in the low temperature range (T < 50 K). Fit curve for the slow component. See also Fig. 31[156)]

of the $1A'_{1u}$ and $1E'_u$ levels, respectively, is described by the following simplified rate equations:

$$\frac{dn_2}{dt} = -[k_2^{em} + k_2^{r\ell} + k_{21}^{r\ell}] \cdot n_2 + k_{12}^{r\ell} \cdot e^{-\Delta E/kT} \cdot n_1 + R_2(t) \quad (18)$$

$$\frac{dn_1}{dt} = -[k_1^{em\|} + k_1^{em\perp} + k_1^{r\ell} + k_{12}^{r\ell} \cdot e^{-\Delta E/kT}] \cdot n_1 + k_{21}^{r\ell} \cdot n_2 \quad (19)$$

The feeding rate $R_2(t)$ does not influence the decay of $n_1(t)$ and $n_2(t)$ for times longer than 1 ns after the excitation pulse, since the $2A'_{2u}$ lifetime ($\tau_{2A'_{2u}} \lesssim 0.3$ ns) is extremely short compared to the low-temperature lifetime of $1E'_u$ ($\tau_{1E'_u} \approx 0.5$ μs) and $1A'_{1u}$ ($\tau_{1A'_{1u}} \approx$ 5 ms). From a detailed discussion of the relative magnitudes of the involved values of the different rates a fit curve of the deactivation rate for YbCP at low temperatures results, as shown in Fig. 32.

F. Polarized Spectroscopy Under High Pressure

The optical transition energies of the crystalline tetracyanoplatinates(II) strongly depend on the metal-metal distance R within the chains. As shown in Sect. E the transition energies can be "adjusted" from the UV over the whole visible spectral range to the near infrared by R-variation through chemical substitution of the cations. This adjustment, however, is *discontinuous*. From the knowledge of the compressibility in chain direction[164, 165] it is expected that application of high pressure allows to reduce the R values, and thus to shift the transition energies *continuously* between the values belonging to the different cations at normal pressure, and furthermore, to reduce R far down to Pt–Pt distances which do not exist under ambient conditions. The controlled variation of the electronic transitions over a range of many thousands of wavenumbers yields valuable additional information about the nature of the electronic states. Moreover, by these distinct properties the MCP compounds represent interesting model systems.

F.I. Pressure Dependence of Emission Spectra

The pressure dependence of the unpolarized emission is reported in Refs. 164, 166. Polarized emission spectra of BaCP[133] at several pressures are shown in Fig. 33. A series of MCP compounds with other cations and thus with other R-values has also been investigated (see Table 9).

Tuning of the Transition Energies

The Figs. 34 to 37 show the pressure dependence of the energies of the peak maxima of BaCP, (enH_2)CP, YCP, and MgCP. With increasing pressure the emission peaks are strongly shifted to the red. The values of the slopes in the low-pressure region ($p \lesssim$ 5–10 kbar) are reproduced in Table 9. (These are the largest values reported in the

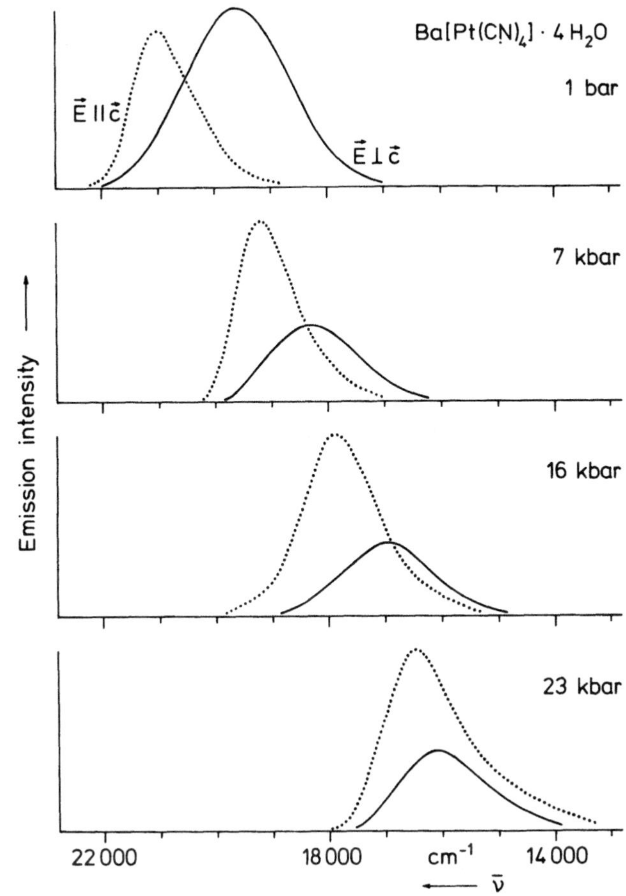

Fig. 33. Polarized emission spectra of single-crystal $Ba[Pt(CN)_4] \cdot 4\ H_2O$ at different hydrostatic pressures (T = 295 K)[133]. The emission intensities at different pressures cannot be compared. The excitation wavelength was varied with pressure to fit approximately the maximum of the $\vec{E}\|\vec{c}$ polarized reflectance. For the high pressure investigations a modified sapphire cell of Bridgman's opposed anvil type was used. The pressure was determined by the amount of red-shift of the R_1- and R_2-lines[167] of ruby crystals placed around the sample

literature for solid-state compounds.) Using different MCP compounds, for example NaCP, CaCP, and MgCP, it is possible to tune the emission energies from the UV [$\bar{\nu}_{max}(\vec{E}\|\vec{c}$, NaCP, p = 1 bar) = 26,750 cm^{-1}] to the IR [$\bar{\nu}_{max}(\vec{E}\|\vec{c}$, MgCP, p = 20 kbar) = 11,200 cm^{-1}] (Fig. 38[113, 114]). This strong pressure dependence allows to study interesting physical properties, such as tunable energy transfer (see Sect. F.II.), tuning off of self-traps (see Sect. D.II.), pressure-induced phase transitions (see below), and pressure- or energy gap-dependent properties of the radiationless relaxation. The emission intensity is reduced by one to two orders of magnitude within the applied pressure range[68].

Transformation Regions and Phase Transitions

Most of the investigated MCP compounds exhibit ranges where pressure increase of several kbar does not shift the emission peak energies[37, 113, 114, 133, 143, 168, 171] (see the

Table 9. Pressure dependent emission properties of different tetracyanoplatinates(II) at 295 K

Compound	$\bar{\nu}_{max}$(1 bar) [cm^{-1}]		$\bar{\nu}_{max}$(20 kbar) [cm^{-1}]		$(\Delta\bar{\nu}/\Delta p)_{p=1\,bar}$ [cm^{-1}/kbar]		Ref.	Remarks
	$\vec{E}\parallel\vec{c}$	$\vec{E}\perp\vec{c}$	$\vec{E}\parallel\vec{c}$	$\vec{E}\perp\vec{c}$	$\vec{E}\parallel\vec{c}$	$\vec{E}\perp\vec{c}$		
$Na_2[Pt(CN)_4]\cdot 3\,H_2O$	26,750	24,400	24,840	22,600	−220	−150	114	10–20 kbar[a]
$Cs_2[Pt(CN)_4]\cdot 1\,H_2O$	22,500[b]		20,300[b]		−115[b]		168	
$K_2[Pt(CN)_4]\cdot 3\,H_2O$	22,900[b]				−130[b]		164, 166	
$Ca[Pt(CN)_4]\cdot 5\,H_2O$	22,250	20,700	19,000	18,200	−200	−140	114	18–23 kbar[a]
$Eu_2[Pt(CN)_4]_3\cdot 18\,H_2O$	22,000[c]	19,000[c]	17,900	16,300			169	
$Sm_2[Pt(CN)_4]_3\cdot 18\,H_2O$	21,500	19,700	18,400	17,700	−260	−180	143	8–14 kbar[a]
$CaCs_2[Pt(CN)_4]_2\cdot n\,H_2O$	21,000[b]		19,200[b]		−90[b]		168	
$Ba[Pt(CN)_4]\cdot 4\,H_2O$	21,000	19,500	17,200	16,800	−270	−200	133	6–12 kbar[a]
	19,800[b]				−167[b]		164, 166	
$Li_2[Pt(CN)_4]\cdot 4\,H_2O$	21,050	19,600	14,750	14,500	−325	−267	36	–[d]
$(enH_2)[Pt(CN)_4]\cdot 0.5\,H_2O$	19,700	18,550	15,500	–[e]	−363	−288	37	10–17 kbar[a]
$KNa[Pt(CN)_4]\cdot 3\,H_2O$	19,500	18,500	16,100	–[e]	−210	−120	168	11–16 kbar
$KLi[Pt(CN)_4]\cdot 2\,H_2O$	18,500	17,800	13,800	–[e]	−320	−220	168	–[d]
$Y_2[Pt(CN)_4]_3\cdot 21\,H_2O$	17,800	16,950			−200	−160	144	
					−400[f]	−270[f]	144	P_{trans} = 5 kbar[g]
$Mg[Pt(CN)_4]\cdot 7\,H_2O$	17,550	16,800	12,850	–[e]	−320	−280	115	–[d]
	16,700[b]				−320[b]		164, 166	

[a] Transformation region; [b] Unpolarized emission; [c] Energies uncertain; [d] No transformation region detected; [e] No $\vec{E}\perp\vec{c}$ polarized emission detectable; [f] At p > 5 kbar; [g] Phase transition

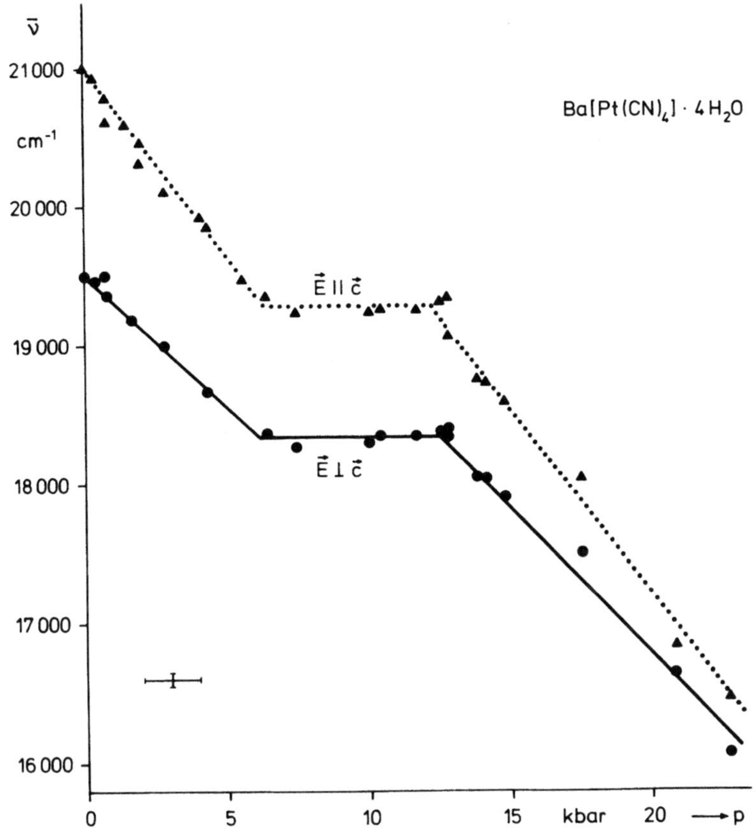

Fig. 34. Energies of the emission peak maxima of single-crystal $Ba[Pt(CN)_4] \cdot 4\,H_2O$ versus pressure at 295 K[133]

Figs. 34, 35 and 38). Since at normal pressure the emission energies are very sensitive to a change of the R value, it is assumed that the intrachain Pt–Pt distances do not change appreciably in these pressure regions. It is likely, that structural transformations occur. For example, rearrangements of hydrogen bonds which link the nitrogens of the $[Pt(CN)_4]^{2-}$ complexes to the surroundings might result within the transformation range. The occurrence of structural transformations is corroborated by the observation that the crystals sometimes crack in these transformation ranges, accompanied by a loss of their anisotropic emission properties[170, 171]. The reduced red-shift for MgCP above 20 kbar can be related to a decrease of the intrachain compressibility[114, 164]. YCP and presumably most compounds of the type $M_2^{III}[Pt(CN)_4]_3 \cdot 21\,H_2O$ show a very different behavior under pressure[43, 130, 144]. YCP exhibits the usual red-shift for the differently polarized transitions up to about 5 kbar. At $p_{trans} \approx 5$ kbar the emission spectra are discontinuously red-shifted (by 750 cm^{-1} for $\vec{E} \| \vec{c}$), cf. Fig. 36. This behavior is explained as a first-order phase transition. Applying Eq. (7) the value for the shrinkage of R is calculated to $\Delta R = (-0.03 \pm 0.01)$ Å. Further pressure increase again leads to a continuous red-shift with the largest slope which has been observed up to now ($\Delta\bar{\nu}/\Delta p = -400$ cm^{-1}/kbar for $p > 5$ kbar). The process of the pressure-induced phase transition is reversible for sev-

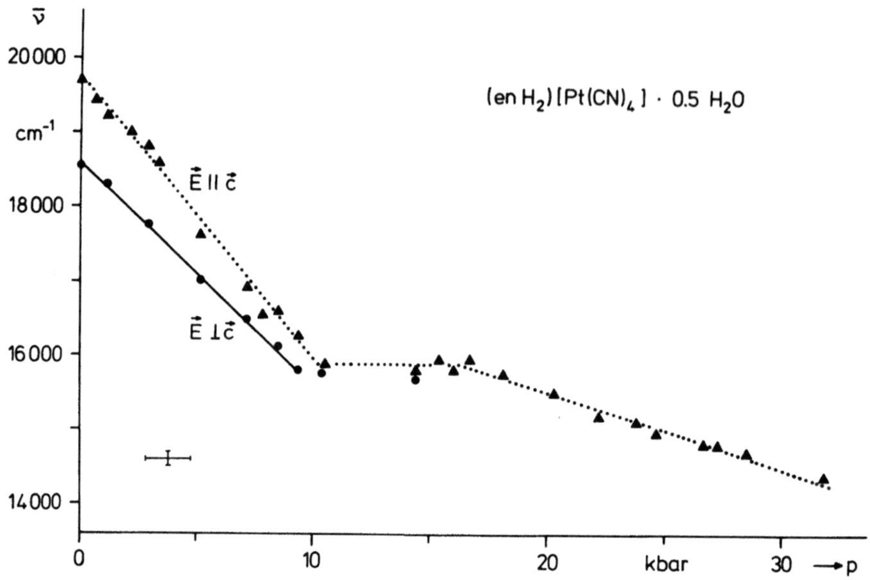

Fig. 35. Energies of the emission peak maxima of single-crystal $(enH_2)[Pt(CN)_4] \cdot 0.5\ H_2O$ versus pressure at 295 K[37]

eral times. It may be mentioned that YCP and a series of other compounds of the same stoichiometry also undergo a first order phase transition by temperature reduction[43, 130].

Cation Substitution and High Pressure Application

The Pt–Pt distance R in chain direction can be varied by cation substitution which leads to the empirical Eq. (7) for the $\vec{E}\|\vec{c}$ polarized emission maxima. Using this relation and

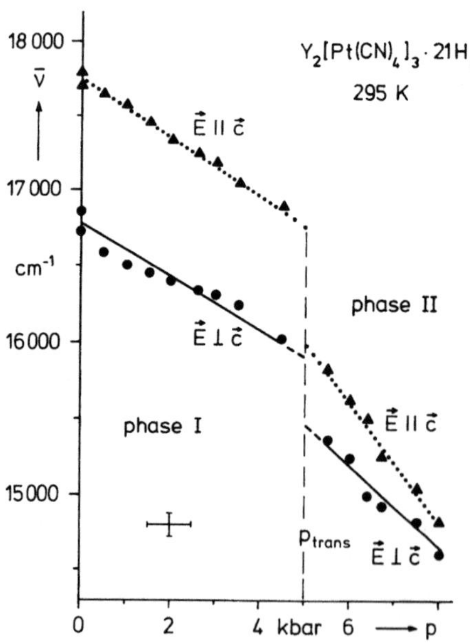

Fig. 36. Energies of the emission peak maxima of single-crystal $Y_2[Pt(CN)_4]_3 \cdot 21\ H_2O$ versus pressure at 295 K[144] (see also Ref. 130)

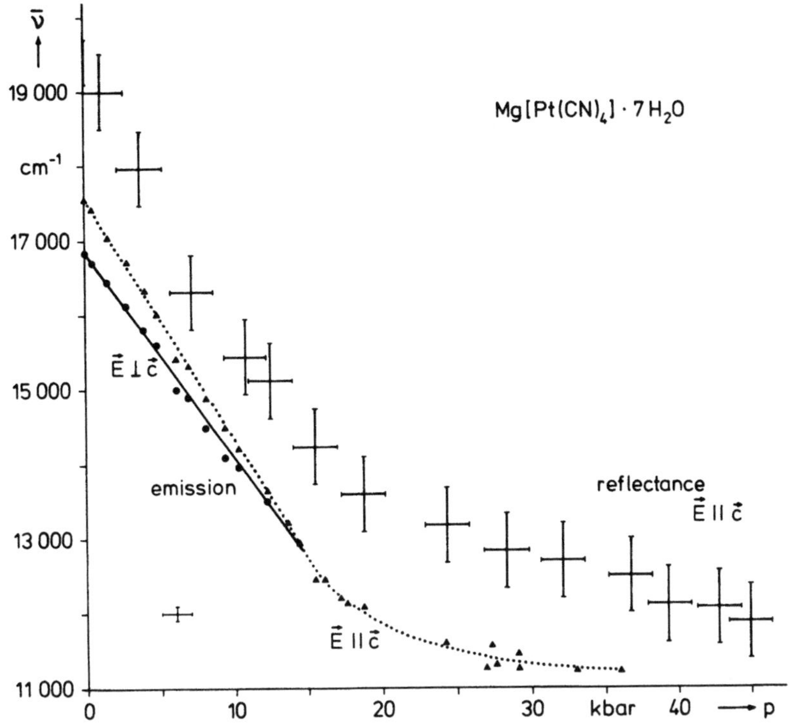

Fig. 37. Energies of the emission and reflectance peak maxima of single-crystal $Mg[Pt(CN)_4] \cdot 7\,H_2O$ versus pressure at 295 K. The cross at the left hand side represents the experimental uncertainty of the emission maxima[114, 115, 170)]

the linear compressibility $\varkappa_c$ in chain direction, $\varkappa_c = -(dR/dp)/R$, the slope of the pressure dependence of the emission energies can be calculated[115)]. For example, the pressure induced red-shift of the $\vec{E}\|\vec{c}$ polarized emission peak energy for MgCP near $R_0 = 3.155$ Å, calculated with the empirical value of the compressibility (from Fig. 39 and Ref. 164), amounts to

$$\frac{d\bar{\nu}_{em}}{dp}(\vec{E}\|\vec{c}) = -(330 \pm 40)\ \mathrm{cm^{-1}/kbar} \tag{20}$$

This is in good agreement with the experimental result (see Table 9),

$$\frac{\Delta\bar{\nu}_{em}}{\Delta p}(\vec{E}\|\vec{c}) = -(320 \pm 20)\ \mathrm{cm^{-1}/kbar} \tag{21}$$

Good agreement also is found for BaCP. The linear compressibilities of several tetracyanoplatinates calculated from the pressure dependence of the $\vec{E}\|\vec{c}$ emission energies are summarized in Table 10 and compared with data from X-ray investigations.

The agreement of the corresponding compressibility values allows to assume that Eq. (7) is valid also for high pressures and the correlated Pt–Pt distances R. Further, the

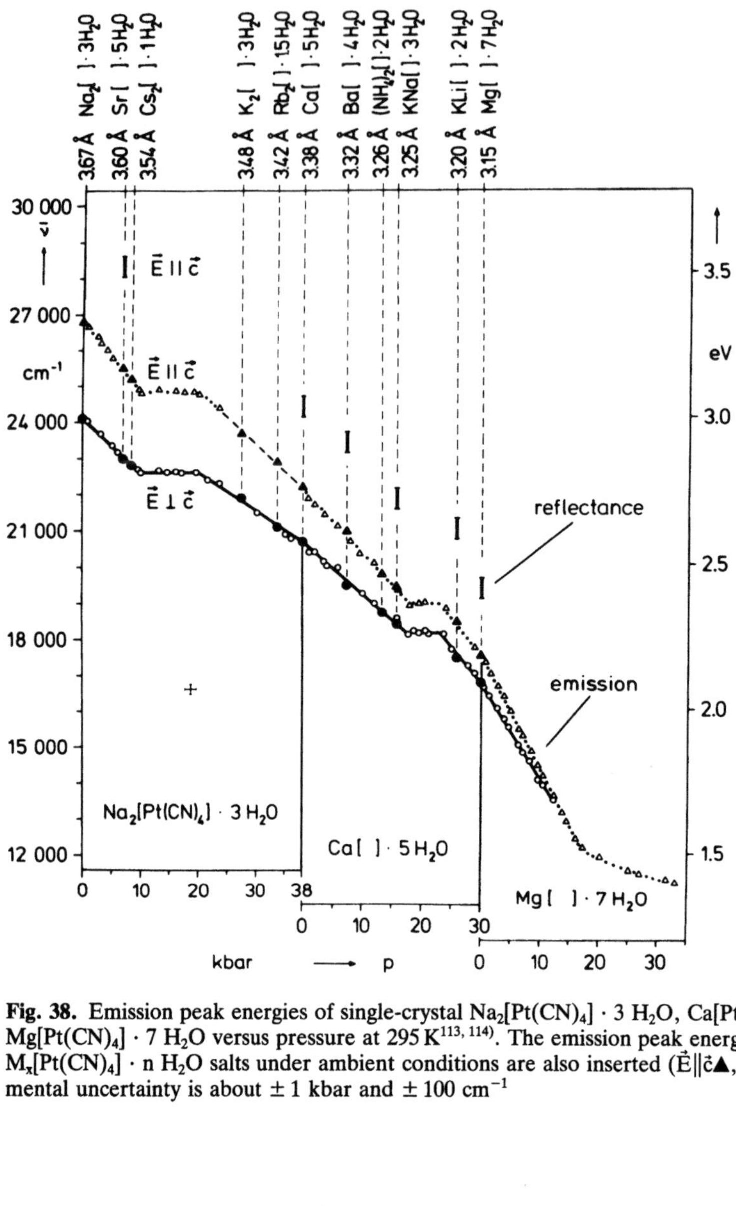

Fig. 38. Emission peak energies of single-crystal $Na_2[Pt(CN)_4] \cdot 3\ H_2O$, $Ca[Pt(CN)_4] \cdot 5\ H_2O$, and $Mg[Pt(CN)_4] \cdot 7\ H_2O$ versus pressure at 295 K[113, 114]. The emission peak energies for various other $M_x[Pt(CN)_4] \cdot n\ H_2O$ salts under ambient conditions are also inserted ($\vec{E}\|\vec{c}$▲, $\vec{E}\perp\vec{c}$●). The experimental uncertainty is about ± 1 kbar and ± 100 cm^{-1}

pressure induced reduction of the intra-complex Pt–C distances and of the inter-chain distances obviously exerts no dominant influence on the energy of the $\vec{E}\|\vec{c}$ emission.

The $\vec{E}\perp\vec{c}$ polarized emission data may also be used to calculate the linear compressibility. However, this seems to be correct only for low pressures[114, 115]. The restriction upon low pressures may be due to the fact that the $\vec{E}\perp\vec{c}$ emission results from self-trapped states. Compared with the $\vec{E}\|\vec{c}$ emitting state the energy of the self-trapped state shows a weaker pressure dependence. This pressure-induced relative energy increase of the self-trapped state is composed of a reduction of the trapping stabilization[111] and of an increase of the ligand field splitting energy of the $[Pt(CN)_4]^{2-}$ ions[174].

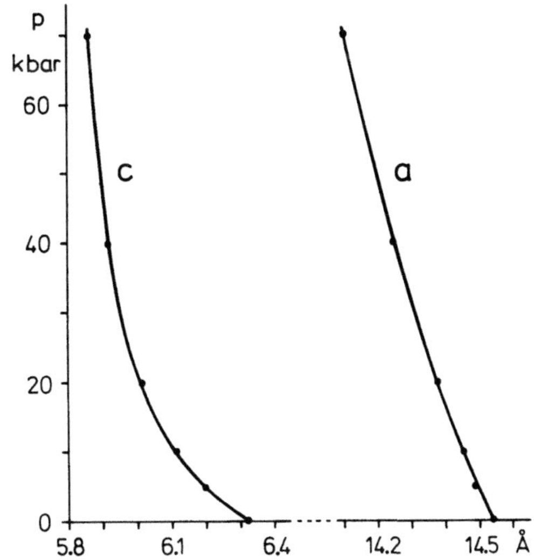

Fig. 39. Lattice constants a and c ≡ 2 R for $Mg[Pt(CN)_4] \cdot 7\,H_2O$ versus pressure at 295 K[164)]

Mean Thermal Expansion

Using the correct Pt–Pt distances R(T) or R(p) it is possible to determine the pressure dependence of the coefficient of thermal expansion $\alpha_c(p)$. According to Eq. (11) the mean linear coefficient of thermal expansion (in K^{-1}) is given by:

$$\bar{\alpha}_c(p) = \frac{1}{R}\frac{\Delta R}{\Delta T} = \frac{\bar{R}^3}{3 \cdot 8.0 \cdot 10^5} \frac{\Delta\bar{\nu}(p, \Delta T)}{\Delta T} \qquad (22)$$

Table 10. Linear compressibility for different tetracyanoplatinates at 295 K and p = 1 bar

Compound	$\varkappa_c$ (emission)[a] [$kbar^{-1}$]	$\varkappa_c$ (X-ray) [$kbar^{-1}$]
$Na_2[Pt(CN)_4] \cdot 3\,H_2O$	4.5×10^{-3}	
$K_2[Pt(CN)_4] \cdot 3\,H_2O$	2.3×10^{-3}	
$Ca[Pt(CN)_4] \cdot 5\,H_2O$	3.2×10^{-3}	
$Sm_2[Pt(CN)_4]_3 \cdot 18\,H_2O$	4.1×10^{-3}	
$Ba[Pt(CN)_4] \cdot 4\,H_2O$	4.1×10^{-3}	4.2×10^{-3}, Ref. 165
$Li_2[Pt(CN)_4] \cdot 4\,H_2O$	5.0×10^{-3}	
$(enH_2)[Pt(CN)_4] \cdot 0.5\,H_2O$	5.6×10^{-3}	
$KNa[Pt(CN)_4] \cdot 3\,H_2O$	3.0×10^{-3}	
$KLi[Pt(CN)_4] \cdot 2\,H_2O$	4.4×10^{-3}	
$Y_2[Pt(CN)_4]_3 \cdot 21\,H_2O$	2.7×10^{-3} 5.4×10^{-3} [b]	
$Mg[Pt(CN)_4] \cdot 7\,H_2O$	4.2×10^{-3}	4.3×10^{-3}, Ref. 115, 164 Fig. 39
$K_2[Pt(CN)_4]Br_{0.3} \cdot 3\,H_2O$	–	1.7×10^{-3}, Ref. 172, 173

[a] Calculated from spectroscopic data using Eq. (7) and the data of Table 9 (experimental error ± 15%); [b] For p > 5 kbar

$\overline{R}$ (in Å) is the mean Pt–Pt distance. $\Delta\bar{\nu}(p, \Delta T)$ (in cm^{-1}) is the red-shift of the $\vec{E}\|\vec{c}$ polarized transition resulting from the temperature variation ΔT. For example, for MgCP one gets from Fig. 40 the values

$$\Delta\bar{\nu}(p, \Delta T) = \{\bar{\nu}_{max}(\vec{E}\|\vec{c}, 295\ K) - \bar{\nu}_{max}(\vec{E}\|\vec{c}, 110\ K)\} \tag{23}$$

Figure 41 shows the resulting pressure dependence of $\Delta\bar{\nu}(p, \Delta T)$ (left scale). It gives also the calculated values $\bar{\alpha}_c(p)$ (right scale) with $\Delta T = 295\ K - 110\ K = 185\ K$ and $\overline{R} = 3.07$ Å. $\bar{\alpha}_c$ is strongly pressure dependent and decreases by a factor of at least 10 to 15 up to 15 kbar[170].

Independently, the linear thermal expansion coefficient can be expressed in a non-quantummechanical approach to[175]

$$\bar{\alpha}_c = \frac{3}{4}\frac{k_B}{R}\frac{A}{C^2} \tag{24}$$

k_B is the Boltzmann constant, A is the anharmonic and C the harmonic coefficient of the intermolecular potential energy in direction of the crystallographic c-axis. The pressure

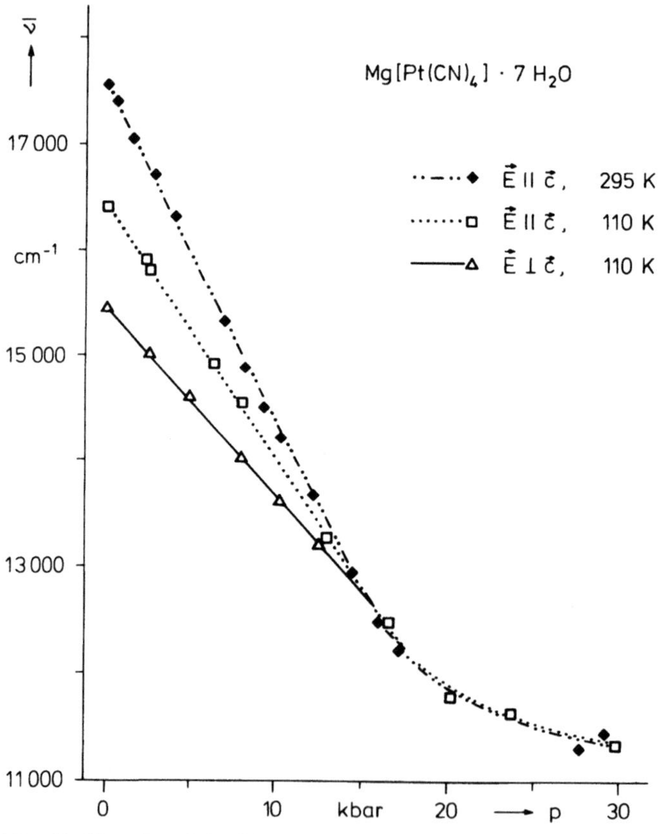

Fig. 40. Energies of the differently polarized emission peak maxima for single crystal $Mg[Pt(CN)_4] \cdot 7\,H_2O$ at different temperatures versus pressure. The data at T = 110 K (± 10 K) result from $Mg[Pt(CN)_4] \cdot 7\,D_2O$[114, 115]

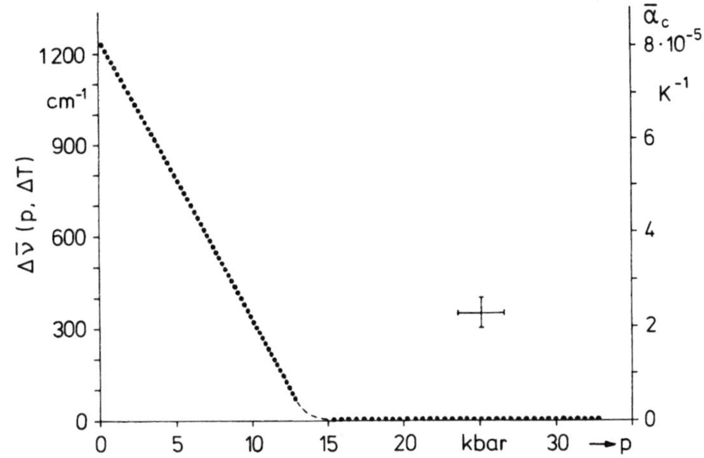

Fig. 41. Red-shift $\Delta\bar{\nu}(p, \Delta T)$ of the $\vec{E}\|\vec{c}$ polarized emission for a temperature reduction of $\Delta T = (295\ K - 110\ K) = 185\ K$ (left scale). The 295 K and the 110 K spectra are recorded for single crystal $Mg[Pt(CN)_4] \cdot 7\ H_2O$ and $\cdot\ 7\ D_2O$, respectively. The right hand scale gives the mean linear coefficient of thermal expansion, calculated using Eq. (22)[170]

dependence of the harmonic term C goes like c_{33}, which increases from 1 bar to 15 kbar by a factor of three to four[111]. Thus, the reduction of $\bar{\alpha}_c$ is mainly determined by the strong increase of the harmonic term[170].

The strong pressure-dependent increase of stiffness has an interesting consequence for the intra-chain coupling of the tetracyanoplatinates with small Pt–Pt distances. For example, near 15 kbar (T = 295 K) MgCP becomes so rigid that the effect of both self-trapping and thermal expansion could not be registered any longer.

F.II. Pressure-Induced Tuning of Radiationless Energy Transfer

Non-radiative energy transfer has been observed for a series of compounds of the type $Ln_2[Pt(CN)_4]_3$ with Ln = Pr, Nd, Sm, Eu, Ho, Er, Tm[30, 93, 95, 113, 142, 143, 169, 176] (see also Table 11). In these compounds the tetracyanoplatinate(II) chains (donors) are excited and the energy is transferred to the rare earth ions (acceptors) under condition of a spectral overlap between the $\vec{E}\perp\vec{c}$ polarized chain-emission and a Ln-acceptor absorption. Since high pressure allows to shift the donor emission over several thousands of wavenumbers the spectral overlap can be tuned in a wide range and thus the energy transfer is strongly pressure dependent. Figure 42 shows, as an example, the results obtained for EuCP[169]. At normal pressure for this compound the spectral overlap integral of the $\vec{E}\perp\vec{c}$ emission and the acceptor $^5D_1 \leftrightarrow {^7F_0}$ transition does not vanish. An emission from $^5D_1$ and $^5D_0$ is observed and the $\vec{E}\perp\vec{c}$ polarized donor component is strongly quenched compared for example to LaCP. At about 20 kbar the $\vec{E}\perp\vec{c}$ chain emission is red-shifted by 2200 cm^{-1}. Now the spectral overlap integral of the chain emission with $^5D_1 \leftrightarrow {^7F_0}$ is zero but non-zero with $^5D_0 \leftrightarrow {^7F_0}$ and, thus, only emission from $^5D_0$ is found. At $p \approx 30$ kbar the line structure resulting from Eu^{3+} is almost quenched completely, since the spectral overlap of the $\vec{E}\perp\vec{c}$ emission with all possible

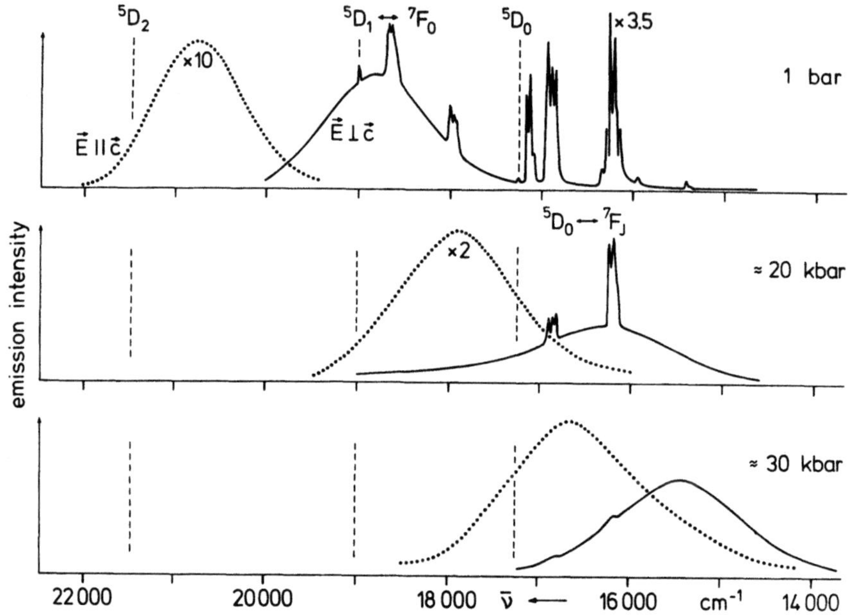

Fig. 42. Polarized emission spectra of $Eu_2[Pt(CN)_4]_3 \cdot 18\ H_2O$ single crystals at different pressures at $T = (100 \pm 10)$ K[169]. (The $\vec{E}\|\vec{c}$ polarized emission intensities of the spectra have to be multiplied by the given factors.)

acceptor transitions is drastically reduced. It is emphasized that the $\vec{E}\perp\vec{c}$ emitting states act *selectively* as donor states while the high energetic $\vec{E}\|\vec{c}$ emitting states yield no remarkable energy transfer. This presumably results from the fact that the emission decay rate of the highly allowed $\vec{E}\|\vec{c}$ transition ($\tau(\vec{E}\|\vec{c}) < 300$ ps[142]) is higher than the rate of energy transfer. Similar results are found for SmCP[143]. The mechanism of the energy transfer can be understood on the basis of an exchange type of transfer according to the Förster-Dexter mechanism[177].

F.III. Triboluminescence

Triboluminescence (TL) is the emission of light caused by the application of mechanical stress on solids. Current research is concentrated upon the emitting excited states responsible for TL, the fracture mechanism giving rise to TL and the structural properties required for TL activity[178, 179]. Sensitive probes for studying the pressure at the TL-active sites are the tetracyanoplatinates because of the large pressure-dependent shift of their luminescence maxima, see Sect. F.I. As found for several tetracyanoplatinates[180] the positions of the peak maxima in the triboluminescence experiments are the same (within experimental error) as those in the 1 bar photoluminescence experiments. Thus, the TL takes place at sites where the stress is relaxed and not at the tips of cracks under high-pressure conditions. These results, together with the good correlation between polar space group and TL activity support the TL excitation mechanism based on electrical charging of newly created fracture surfaces.

G. Magnetic Field Effects

The investigation of magnetic field effects on the luminescence of MCP crystals provides further understanding about the emitting states, especially the lowest ones, confirming that there is an analogy between the self-trapped exciton states and the lowest excited molecular states of a $[Pt(CN)_4]^{2-}$ ion embedded in the crystal. The magnetic field effects will be reported in separate sections for MCP (M ≠ Ln), for the rare earth salts LnCP, and for mixed crystals MCP with Ni doping.

G.I. *Magnetic Field Effects for $M_x[Pt(CN)_4] \cdot n\, H_2O$ With M ≠ Ln*

The behavior of the low temperature $\vec{E}\perp\vec{c}$ polarized emission spectrum of a typical MCP salt, BaCP, for different magnetic field strengths is presented in Fig. 43[94]. The vector orientation is $\vec{H}\perp\vec{c}$, $\vec{H}\perp\vec{O}$ and $\vec{O}\perp\vec{c}$, where $\vec{O}$ denotes the direction of observation. A relatively large blue shift of the peak maximum of the $\vec{E}\perp\vec{c}$ emission by $\sim 270\ cm^{-1}$ (Fig. 44) and a drastic increase of the intensity (Fig. 45) are observed as the magnetic field is raised from 0 to 6 T. Above ≈ 3 T the spectral position of the $\vec{E}\perp\vec{c}$ emission remains nearly constant, whereas no saturation effect of the intensity growth is found up to 6 T. If the excitation wavelength is changed from 364 to 458 nm the increase in intensity is lowered by a factor of 3.

In the range between 0 and 6 T at 1.9 K the lifetime of the $\vec{E}\perp\vec{c}$ emission decreases continuously from 4.3 ms to 20 μs. The decay of the $\vec{E}\perp\vec{c}$ emission remains mono-exponential over the whole emission band. In Fig. 46 the difference of the reciprocal values of the lifetime at H and H = 0, $1/\tau_H - 1/\tau_0$, is plotted versus the magnetic field strength H on

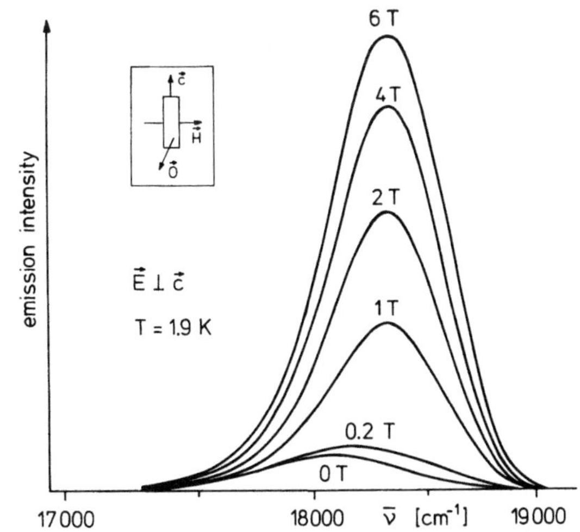

Fig. 43. $\vec{E}\perp\vec{c}$ polarized emission spectrum of single-crystal $Ba[Pt(CN)_4] \cdot 4\ H_2O$ at T = 1.9 K for different magnetic field strengths. Vector orientations $\vec{H}\perp\vec{c}$, $\vec{H}\perp\vec{O}$, $\vec{O}\perp\vec{c}$. $\vec{O}$: direction of observation. (λ_{ex} = 364 nm)[94]

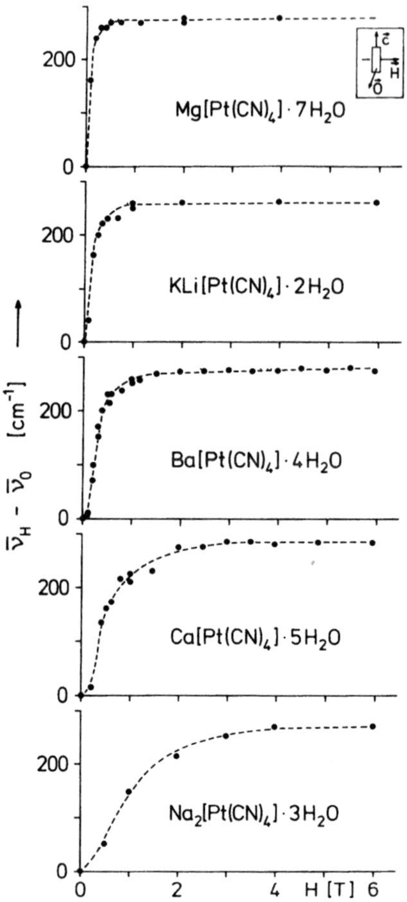

Fig. 44. Blue shift of the $\vec{E}\perp\vec{c}$ polarized emission maximum of several $M_x[Pt(CN)_4] \cdot n\, H_2O$ single crystals as a function of magnetic field strengths. $\bar{\nu}_H$ and $\bar{\nu}_0$ are the wavenumbers of the emission maxima at magnetic field strength H and H = 0, respectively[94]. Vector orientations $\vec{H}\perp\vec{c}$, $\vec{H}\perp\vec{O}$, $\vec{O}\perp\vec{c}$. T = 1.9 K

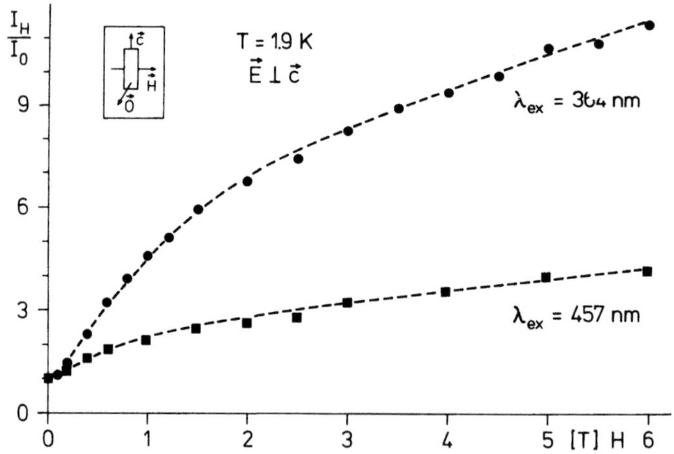

Fig. 45. Relative intensity increase of the $\vec{E}\perp\vec{c}$ polarized emission of $Ba[Pt(CN)_4] \cdot 4\, H_2O$ as a function of magnetic field strengths. I_H and I_0 are the emission intensities at magnetic fields H and H = 0, respectively. Vector orientations $\vec{H}\perp\vec{c}$, $\vec{H}\perp\vec{O}$, $\vec{O}\perp\vec{c}$. Excitation wavelengths: ● ≙ 364 nm, ■ ≙ 457 nm. T = 1.9 K[94]

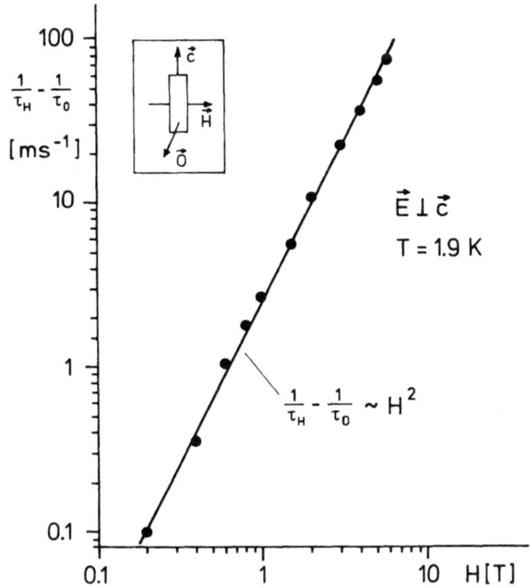

Fig. 46. Magnetic field dependent increase of the $\vec{E}\perp\vec{c}$ deactivation rate of $Ba[Pt(CN)_4] \cdot 4\ H_2O$ at $T = 1.9$ K. τ_H and τ_0 are the emission lifetimes at magnetic field strengths H and H = 0, respectively[94)]

a log-log scale. All measured values lie on a straight line with slope 2, which means that the magnetic field induced change of the deactivation rate of the emitting level is a purely quadratic function of H over a range of more than three orders of magnitude.

For the same vector orientations $\vec{H}\perp\vec{c}$, $\vec{H}\perp\vec{O}$, $\vec{O}\perp\vec{c}$ the low energetic $\vec{E}\|\vec{c}$ emission (at $\approx 18{,}000\ cm^{-1}$, cf. Fig. 19) shows a similar field dependence of its deactivation rate but exhibits a relatively weak magnetically induced intensity growth.

With $\vec{H}\perp\vec{c}$, $\vec{H}\|\vec{O}$, $\vec{O}\perp\vec{c}$ the intensity of the low energetic $\vec{E}\|\vec{c}$ emission increases whereas for $\vec{E}\perp\vec{c}$ an intensity decrease is observed with increasing $\vec{H}$. The blue shift of the emission and the change in the deactivation rate have nearly the same values as for the vector orientations $\vec{H}\perp\vec{c}$, $\vec{H}\perp\vec{O}$, $\vec{O}\perp\vec{c}$.

Much weaker magnetic field effects are observed, when the magnetic field vector is parallel to the c axis of the crystal[151)].

Similar results were obtained for other crystals with different Pt–Pt distances (MgCP, KLiCP, CaCP, NaCP)[94)]. However, a systematic difference between these crystals is obvious. The smaller the intrachain Pt–Pt distance the lower is the magnetic field strength at which the saturation of the blue shift occurs. The maximum of the blue shift, however, does not depend on the Pt–Pt distance (Fig. 44).

The interpretation of the described magnetic field effects can be given within the model of localized, self-trapped excitons (STE). A simplified energy level scheme is shown in Fig. 47[94)]. For abbreviation the prefix 1 at the symbols of the low lying self-trapped states A'_{1u} and E'_u will be omitted below (see also Fig. 13).

As shown in chapter E the $\vec{E}\perp\vec{c}$ polarized emission at H = 0 is allowed for the transition $E'_u \leftrightarrow A'_{1g}$. The transition from the excited A'_{1u} into the ground state, however, requires the vibronic coupling with a ε_g vibration.

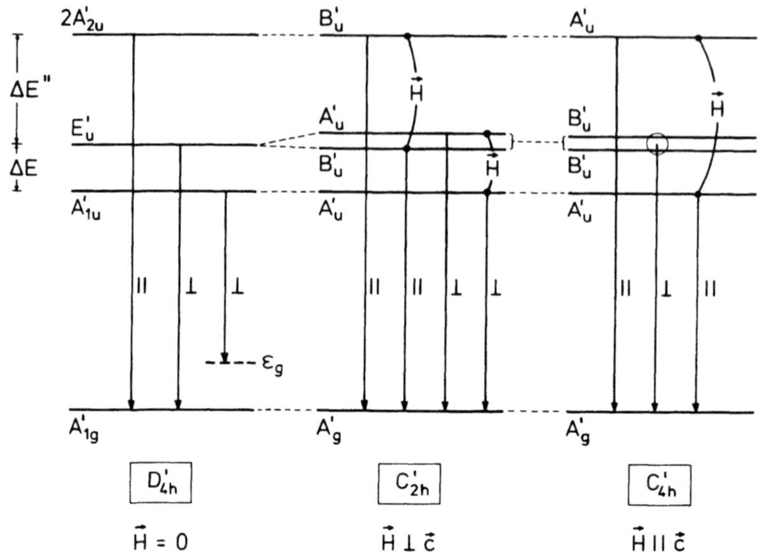

Fig. 47. Radiative transitions from the lowest emitting states of a tetracyanoplatinate(II) crystal *(simplified)*. D'_{4h}: zero magnetic field; C'_{2h}: $\vec{H} \perp \vec{c}$; C'_{4h}: $\vec{H} \| \vec{c}$ (For the values ΔE and $\Delta E''$ see Table 6.)

When the magnetic field is applied in a direction perpendicular to the fourfold symmetry axis ($\vec{H} \perp \vec{c}$) the symmetry is lowered from D'_{4h} to C'_{2h}. Thus, the degeneracy of the E'_u state is removed and one gets three separated levels $A'_u(E'_u)$, $B'_u(E'_u)$, $A'_u(A'_{1u})$ of triplet parentage (Fig. 47). An important consequence of the symmetry lowering is the magnetic field induced mixing of the A'_u states of E'_u and A'_{1u} parentage.

If the A'_u wavefunction components of the (unperturbed) excited states A'_{1u} and E'_u are $|A'_u(A'_{1u})\rangle_0$ and $|A'_u(E'_u)\rangle_0$, respectively, the lower lying A'_u can be approximated by first order perturbation theory:

$$|A'_u(A'_{1u})\rangle \approx |A'_u(A'_{1u})\rangle_0 + \frac{\langle A'_u(E'_u)|\mathcal{H}'|A'_u(A'_{1u})\rangle_0}{E(E'_u) - E(A'_{1u})} \cdot |A'_u(E'_u)\rangle_0 \tag{25}$$

The perturbation operator $\mathcal{H}'$ is proportional to the absolute value of the magnetic field strength $\vec{H}$[75],

$$\mathcal{H}' = g \cdot \vec{H} \cdot (\vec{L} + 2\vec{S}) \sim |\vec{H}| = H\,, \tag{26}$$

Thus, the wavefunction can be written as

$$|A'_u(A'_{1u})\rangle \approx |A'_u(A'_{1u})\rangle_0 + c \cdot H \cdot |A'_u(E'_u)\rangle_0 \tag{27}$$

with

$$c \cdot H = \frac{\langle A'_u(E'_u)|\mathcal{H}'|A'_u(A'_{1u})\rangle_0}{E(E'_u) - E(A'_{1u})} \tag{28}$$

and the electric dipole transition moment of the $A'_u(A'_{1u}) \rightarrow A'_g(A'_{1g})$ emission with $\vec{E}\perp\vec{c}$ polarization is

$$\langle A'_g(A'_{1g})|\mu|A'_u(A'_{1u})\rangle = c \cdot H \cdot \langle A'_g(A'_{1g})|\mu|A'_u(E'_u)\rangle_0 \quad (29)$$

Recalling that the E'_u state exhibits a $\vec{E}\perp\vec{c}$ allowed radiative transition to the A'_{1g} ground state, it follows from Eq. (29) that the mixing of the wave functions opens a new radiative channel from the lowest A'_u directly to the vibrationally non excited A'_g ground state. This explains the large, magnetically induced blue shift of the emission and its saturation behavior at 1.9 K, when with increasing magnetic field finally the emission from the lowest A'_u dominates. The observed value of 270 cm^{-1} fits approximately the energy of the molecular ε_g vibration of the $[Pt(CN)_4]^{2-}$ complex. This interpretation is also supported by the fact that the maximum value of the blue shift is independent of the Pt–Pt distance R.

The arguments given above also allow to explain the observed quadratic H-dependence of the deactivation rate (Fig. 46). From Eq. (29) it follows,

$$1/\tau_H - 1/\tau_0 = \alpha \cdot H^2 \quad (30)$$

From group theoretical selection rules it is predicted that the additional, magnetically induced $\vec{E}\perp\vec{c}$ polarized transition from the A'_u to the A'_g is not observable for the vector orientation $\vec{H}\perp\vec{c}$, $\vec{H}\|\vec{O}$, $\vec{O}\perp\vec{c}$. (The relevant transition matrix element is zero, when the wave vector of the photon is parallel to the vector $\vec{H}$.) This is in agreement with the observed relative intensity decrease in $\vec{E}\perp\vec{c}$, when the field is raised[94]. (Intensity changes found for $\vec{H}\|\vec{O}$ are at least partially due to a Faraday effect, which has to be considered for this vector orientation[181].)

When the magnetic field is applied parallel to the c axis, $\vec{H}\|\vec{c}$, the symmetry changes from D'_{4h} to C'_{4h}. In this case the A'_{1u} mixes with the A'_{2u} (Fig. 47), which opens up a new, $\vec{E}\|\vec{c}$ polarized radiative channel from A'_u to the A'_g. Because of the relatively large energetic separation between A'_{2u} and A'_{1u} the magnetic field effects should be much weaker in C'_{4h} than in C'_{2h}, in agreement with the experimental results[94].

G.II. *Magnetic Field Effects for $Ln_2[Pt(CN)_4]_3 \cdot n\, H_2O$*

Properties of the $\vec{E}\perp\vec{c}$ emission of LnCP's under high magnetic fields at T = 1.9 K are given in Table 11[95, 156].

Among the LnCP's only the LuCP salt shows magnetic field effects similar to the MCP's with M ≠ Ln. For the vector orientations $\vec{H}\perp\vec{c}$, $\vec{H}\perp\vec{O}$, $\vec{O}\perp\vec{c}$ the $\vec{E}\perp\vec{c}$ polarized emission of LuCP shows the same intensity increase (factor ≈12 from H = 0 to H = 6 T) but a blue shift of only ≈120 cm^{-1} (at H > 1 T)[95]. The difference to the blue shift of the MCP's with M ≠ Ln ($\Delta\bar{\nu}_{max} \sim 270$ cm^{-1}) may be due to a perturbation of the local D'_{4h} symmetry in LuCP already at H = 0 and, accordingly, to a non-vanishing probability for an optical transition from the parentage term A'_{1u} to the ground state at zero magnetic field (cf. Rev. 156). Further, for LuCP the deactivation rate $1/\tau_H$ is also a quadratic function of the magnetic field strength H. For the vector orientations $\vec{H}\perp\vec{c}$, $\vec{H}\|\vec{O}$ and $\vec{H}\|\vec{c}$, $\vec{H}\perp\vec{O}$ the magnetic field effects of LuCP also harmonize with the effects

Table 11. Properties of the $\vec{E}\perp\vec{c}$ emission of $Ln_2[Pt(CN)_4]_3 \cdot n\,H_2O$ at high magnetic fields at 1.9 K. Vector orientations $\vec{H}\perp\vec{c}$, $\vec{H}\perp\vec{O}$, $\vec{O}\perp\vec{c}$[95)]

$Ln[Pt(CN)_4]_3 \cdot n\,H_2O$		Spectral overlap[b]	H = 0 T τ_0[μs]	H = 6 T τ_{6T}[μs]	τ_0/τ_{6T}	Blue-shift $\Delta\bar{\nu}_{max}$ [cm^{-1}]	I_{6T}/I_0	H = 0 T $\bar{\nu}_{max}$ [cm^{-1}]	H = 6 T $\bar{\nu}_{max}$ [cm^{-1}]	Number of f electrons	Ionization energy/eV $Ln^{2+} \rightarrow Ln^{3+}$
LaCP		–	800[c]	20	40	0	5–6			0	19.18
CeCP		–	90[c]	20	4–5	0	2			1	20.11
PrCP[a]	yellow salts	+			–	0	–			2	21.66
NdCP[a]	n = 18	+	$1\text{–}2 \cdot 10^{-3}$	$1\text{–}2 \cdot 10^{-3}$	–	0	–	18,480 ± 20	18,480 ± 20	3	22.06
SmCP[a]		+			–	0	–			5	23.68
EuCP[a]		+			–	0	–			6	25.13
GdCP		–	90[c]	20	4–5	0	2			7	20.72
TbCP		–	–[d]	–	–	–	–	–	–	8	21.99
DyCP		–	–[d]	–	–	–	–	–	–	9	23.11
HoCP	red salts	+	–[d]	–	–	–	–	–	–	10	23.02
ErCP	n = 21	+	$1.3 \cdot 10^{-3}$	$1.3 \cdot 10^{-3}$	1	0	–			11	22.87
TmCP		+	$1.3 \cdot 10^{-3}$	$1.3 \cdot 10^{-3}$	1	0	–	15,400 ± 20	15,400 ± 20	12	23.89
YbCP		–	1.5	1.5	1	0	5			13	24.96
LuCP		–	7300	15	490	≈120	10–12	15,280 ± 20	15,400 ± 20	14	...

[a] Uncertain spectroscopic results because of damages by phase transitions, cf.[130)]; [b] spectral overlap between $\vec{E}\perp\vec{c}$ emission and Ln^{3+} absorption, cf. Sect. F.II.; [c] decay slightly non-exponential; [d] very weak $\vec{E}\perp\vec{c}$ intensity

reported for the MCP's with M $\neq$ Ln. Thus, the influence of the magnetic field on the LuCP emission can be interpreted within the same model, as is applied to the MCP's (M $\neq$ Ln) in Sect. G.I.

On the contrary the other LnCP's (Ln $\neq$ Lu) exhibit distinctly deviating optical properties for $0 \leq H \lesssim 6$ T. The influence of the magnetic field on lifetime, intensity and spectral position of the LnCP emission is rather weak or even vanishing.

In the presence of energy transfer (salts with non-vanishing spectral overlap between $\vec{E} \perp \vec{c}$ emission and Ln^{3+}-absorption[30], see Table 11) the $\vec{E} \perp \vec{c}$ emission has to be ascribed to the prompt emission from E'_u [93, 95, 156]. In these cases the additional non-radiative deactivation of this level is already so effective that a further radiative deactivation channel, opened by the influence of a magnetic field, proves to be inefficient.

In the field free case the optical behavior of the LnCP's (except Ln = Lu and the salts with non-vanishing spectral overlap) can be interpreted by a modified model involving energetically low lying charge transfer (CT) states created by an electron transfer from a (Pt $5\,d_{z^2}$) orbital to a 4f orbital of the cation[93]. Then the low lying A'_{1u} state can be approximated by a wave function

$$|A'_{1u}\rangle_{CT} \approx |A'_{1u}\rangle_0 + \lambda^{CT} \cdot |CT\rangle_0 \qquad (31)$$

with

$$|A'_{1u}\rangle_0 = \Psi(Pt^{2+}) \cdot \Phi(Ln^{3+}) \,, \qquad (32)$$

$$|CT\rangle_0 = \psi(Pt^{3+}) \cdot \phi(Ln^{2+}) \,, \qquad (33)$$

$$\lambda^{CT} = \frac{\langle CT|V_{eff}|A'_{1u}\rangle_0}{E_{CT} - E_{A'_{1u}}} \qquad (34)$$

Therefore the amount of Ln(4f) admixture to the A'_{1u} is the larger the higher the ionization energy E_{CT} of the reduced cation Ln^{2+}, i.e. the lower the acceptor state. (The exceptional case of LuCP is obvious: The 4f shell of Lu^{3+} is completely filled and a corresponding CT state does not exist.)

When a magnetic field $\vec{H}$ is applied the emitting $A'_u(A'_{1u})$ state additionally mixes with $A'_u(E'_u)$, as shown in Sect. G.I.,

$$|A'_u(A'_{1u})\rangle \approx |A'_{1u}\rangle_0 + \lambda^{CT} \cdot |CT\rangle_0 + c \cdot H \cdot |A'_u(E'_u)\rangle_0 \qquad (35)$$

It follows that the probability of emptying the low lying $A'_u(A'_{1u})$ by radiative and non-radiative processes can be written in the form

$$1/\tau_H = k(H) = k(0) + \delta \cdot H + \alpha \cdot H^2 \qquad (36)$$

k(0) corresponds to the reciprocal value of the excited state lifetime at H = 0 (and T = 1.9 K). A comparison of different systems shows that the coefficient α is nearly independent of the structure and the amount of CT mixing and has the value $\alpha \approx 10^3 \cdot s^{-1} \cdot T^{-2}$. About δ no quantitative information is available, except $\delta \cdot H < k(0) + \alpha \cdot H^2$ [95].

Yb^{2+} has a relatively high ionization energy (24.96 eV, see Table 11). Therefore, a large CT admixture is expected for YbCP resulting in an additional optical $0 \rightarrow 0$ transition from the A'_{1u} to the A'_{1g} state, even at $H = 0$. For this reason the $\vec{E} \perp \vec{c}$ emission maximum of YbCP is already blue shifted at $H = 0$ to the position which results for LuCP at $H > 1$ T. Therefore, no magnetic field effect on the energetical position of the $\vec{E} \perp \vec{c}$ emission can be observed.

In the yellow salts of LaCP, CeCP, GdCP the CT states probably lie relatively high above the emitting states of the $[Pt(CN)_4]^{2-}$ chains, entailing only a weak CT admixture. This is compatible with the relatively long emission lifetimes of CeCP and GdCP ($\tau \approx 90$ μs) and of LaCP ($\tau \approx 800$ μs) compared to YbCP. The Pt–Pt distances R of these yellow salts (Table 1) are similar to that of BaCP. Their $\vec{E} \perp \vec{c}$ emission peak maxima at $H = 0$, however, have values observed for BaCP only at $H > 1.5$ T[94]. Presumably, the CT-admixture anticipates the magnetically induced blue shift in these cases, too – despite their comparably low Ln^{2+} ionization energy.

Doping of the red LuCP salt with impurity cations $Ln^{3+} = Eu^{3+}, Gd^{3+}, Tb^{3+}, Yb^{3+}$
The application of these dopants excludes effects of resonant energy transfer. For an impurity concentration of roughly 10% about every second or third $[Pt(CN)_4]^{2-}$ ion has an impurity cation within its first coordination sphere. In the field free case the $\vec{E} \perp \vec{c}$ emission decay is non-exponential but the decay constant of the long-living component is the same as that of pure LuCP. From these facts a strong localization of the excitons is concluded, favoring the approximative description of the lowest excited states A'_{1u} and E'_u in the tetracyanoplatinate(II) crystals by modified wave functions of a $[Pt(CN)_4]^{2-}$ ion embedded in the crystal[95]. If a magnetic field with vector orientation $\vec{H} \perp \vec{c}$, $\vec{H} \perp \vec{O}$, $\vec{O} \perp \vec{c}$ is applied to the doped LuCP, at $H > 1.5$ T the decay becomes mono-exponential and reaches the value of the pure LuCP at this field strength. This is not only due to the fact that the lifetimes of $[Pt(CN)_4]^{2-}$ ions with and without dopant neighbours approach at high field strengths, but also that the emission of $[Pt(CN)_4]^{2-}$-ions with unperturbed environment (and "normal" magnetic field sensibility) eventually exceeds the emission of $[Pt(CN)_4]^{2-}$ ions with dopant neighbours. Therefore, at higher field strengths the influence of doping on lifetime is not oberservable anymore[95].

G.III. *Magnetic Field Effects for Mixed-Crystals $Ba[Pt_{1-x}Ni_x(CN)_4] \cdot 4\,H_2O$*

The $\vec{E} \perp \vec{c}$ polarized emission of mixed-crystal $Ba[Pt_{1-x}Ni_x(CN)_4] \cdot 4\,H_2O$ with $0.01 \lesssim x \lesssim 0.1$ shows an intensity growing by a factor of 7–10 with increasing magnetic field $\vec{H} \perp \vec{c}$ (between $H = 0$ and $H \approx 5$ T), whereas the blue shifts $\bar{\nu}_H - \bar{\nu}_0$ of the emission maxima come up to saturation values at $H \sim 1$ T being different for the various concentrations x (Fig. 48). In Fig. 49 the saturation values of the blue shift are plotted against $-\log x$. A straight line results, reaching the value of the pure BaCP at $-\log x \approx 2$. For $x = 10^{-1.9}$ (≈ 0.01) the shift amounts to ~ 240 cm^{-1} which is close to the value of ~ 270 cm^{-1} for the pure BaCP ($x = 0$)[92].

Increasing dopant concentration and increasing magnetic field strength yield very similar effects in the $\vec{E} \perp \vec{c}$ polarized emission at low temperatures. As shown above the magnetic field opens a new radiative channel from the low lying A'_{1u} state into the

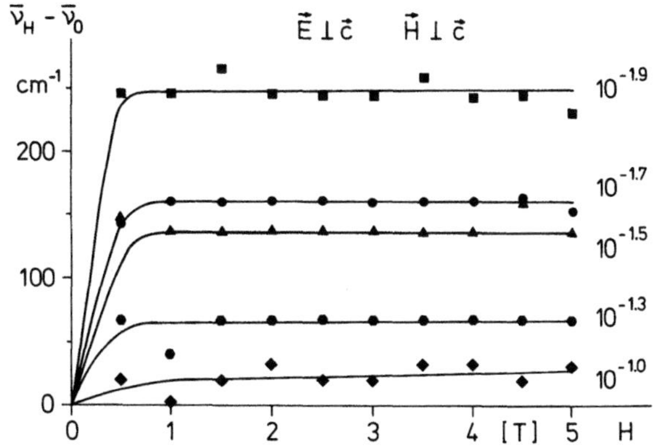

Fig. 48. Blue shift ($\bar{\nu}_H - \bar{\nu}_0$) of the $\vec{E}\perp\vec{c}$ emission maxima of $Ba[Pt_{1-x}Ni_x(CN)_4] \cdot 4\,H_2O$ single crystals for several concentrations x versus magnetic field strength, at 1.7 K[92]. $\bar{\nu}_H$ and $\bar{\nu}_0$ are explained below Fig. 44

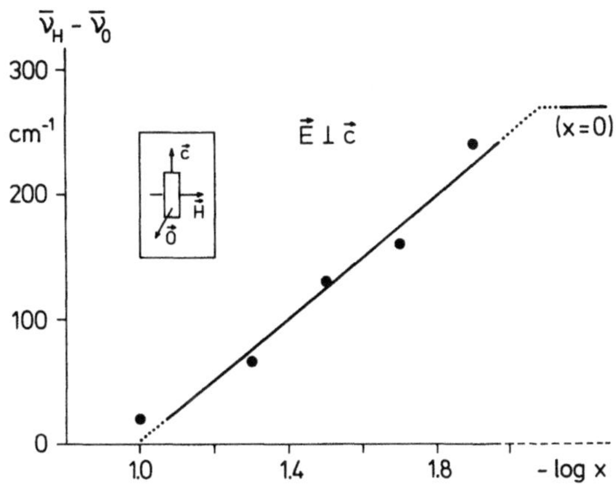

Fig. 49. Saturation values of the blue shift versus $-\log x$ for $Ba[Pt_{1-x}Ni_x(CN)_4] \cdot 4\,H_2O$ single crystals at T = 1.7 K. H > 2 T[92]

vibrationally non-excited ground state A'_{1g}, resulting in a blue shift $\Delta\bar{\nu} \lesssim \bar{\nu}(\varepsilon_g)$. On the other hand, doped $[Ni(CN)_4]^{2-}$ ions depopulate the excited A'_{1u} state by a non-radiative process, enhancing the relative intensity of the emission from the E'_u state[90,91].

Both perturbations, applied magnetic fields and doping, can combine their own blue shifts up to a maximum of $\Delta\bar{\nu} \approx \bar{\nu}(\bar{\varepsilon}_g)$. Therefore the maximum field induced blue shift gets smaller with increasing doping concentration.

Acknowledgments. We would like to thank the Fonds der Chemischen Industrie and the Deutsche Forschungsgemeinschaft for support of our work.

H. References

1. Gmelin, L.: Jahrbuch der Chemie und Physik *36,* 230 (1822)
2. Brewster, D.: Rep. Brit. Assoc. Adv. Sci., 20th Meet., p. 5 (1850)
3. Stokes, G. G.: Philosophical Transactions, Royal Soc. of London *143,* 385 (1853)
4. Yamada, S.: J. Am. Chem. Soc. *73,* 1182 (1951)
5. Krogmann, K.: Angew. Chemie *81,* 10 (1969)
6. Interrante, L. V. (ed.): Extended Interactions between Metal Ions in Transition Metal Complexes, ACS Symp. Series 5, Washington 1974
7. Schuster, H. G. (ed.): One-Dimensional Conductors, Lecture Notes in Physics, Nr. 34, Berlin, Springer-Verlag 1975
8. Keller, H. J. (ed.): Low-Dimensional Cooperative Phenomena. The Possibility of High-Temperature Superconductivity, New York, Plenum Press 1975
9. Keller, H. J. (ed.): Chemistry and Physics of One-Dimensional Metals, New York, Plenum Press 1977
10. Miller, J. S., Epstein, A. J. (ed.): Synthesis and Properties of Low Dimensional Materials, Annals of the New York Academy of Sciences, Vol. *313,* 1978
11. Devresse, J. T., Evrard, R., van Doren, V. E. (ed.): Highly Conducting One-Dimensional Solids, New York, Plenum Press 1979
12. Miller, J. S. (ed.): Extended Linear Chain Compounds, New York – London, Plenum Press, Vol. I–II (1982), Vol. III (1983)
13. Moreau-Colin, M. L.: Structure and Bonding *10,* 167 (1972)
14. Holzapfel, W., Yersin, H., Gliemann, G.: Z. Kristallogr. *157,* 47 (1981)
15. Williams, J. M., Schultz, A. J., Underhill, A. E., Carneiro, K.: in Ref. 12, Vol. I, p. 73
16. Handbook of Chemistry and Physics, 61st ed., p. F-219, 1980–81
17. Ledent, J.: Bull. Soc. Roy. Sci. Liège *41,* 537 (1972)
18. Johnson, P. L., Koch, T. R., Williams, J. M.: Acta Crystallogr. *B33,* 1976 (1977)
19. Dupont, L.: Bull. Soc. Roy. Sci. Liège *36,* 40 (1967); *36,* 471 (1967)
20. Krogmann, K., Stephan, D.: Z. anorg. allg. Chem. *362,* 290 (1968)
21. Brasseur, H., de Rassenfosse, A.: Bull. Soc. Roy. Sci. Liège *5,* 124 (1936)
22. Otto, H. H., Holzapfel, W., Yersin, H., Gliemann, G.: Z. Naturforsch. *31b,* 528 (1976)
23. Otto, H. H., Schulz, H., Thiemann, K. H., Yersin, H., Gliemann, G.: ibid. *32b,* 127 (1977)
24. Johnson, P. L., Koch, T. R., Williams, J. M.: Acta Crystallogr. *B33,* 1293 (1977)
25. Washecheck, D. M., Peterson, S. W., Reis, A. H. jr., Williams, J. M.: Inorg. Chem. *15,* 74 (1976)
26. Dupont, L.: Bull. Soc. Roy. Sci. Liège *38,* 509 (1969)
27. Koch, T. R., Johnson, P. L., Williams, J. M.: Inorg. Chem. *16,* 640 (1977)
28. Fontaine, F.: Bull. Soc. Roy. Sci. Liège *36,* 437 (1968)
29. Holzapfel, W., Yersin, H., Otto, H. H., Gliemann, G.: Ber. Bunsenges. Phys. Chem. *82,* 207 (1978)
30. Yersin, H.: J. Chem. Phys. *68,* 4707 (1978)
31. v. Ammon, W.: Diplomarbeit, University of Regensburg 1977
32. Brasseur, H., de Rassenfosse, A.: Bull. Soc. fr. Minéral. Cristallogr. *61,* 129 (1938)
33. Maffly, R. L., Johnson, P. L., Williams, J. M.: Acta Crystallogr. *B33,* 884 (1977)
34. Holzapfel, W., Gliemann, G.: unpublished results
35. Fontaine, R., Moreau, M. L., Simon, J.: Bull. Soc. fr. Minéral. Cristallogr. *91,* 400 (1968)
36. Daniels, W.: Thesis, University of Regensburg 1982
37. Dillinger, R., Gliemann, G., Pfleger, H. P., Krogmann, K.: Inorg. Chem. *22,* 1366 (1983)
38. Pfleger, H. P.: Thesis, University of Karlsruhe 1981
39. Moreau-Colin, M. L.: Bull. Soc. fr. Minéral. Cristallogr. *91,* 332 (1968)
40. Johnson, P. L., Musselman, R. L., Williams, J. M.: Acta Crystallogr. *B33,* 3155 (1977)
41. Moreau-Colin, M. L.: Bull. Soc. Roy. Sci. Liège *34,* 778 (1965)
42. Beno, M. A., Rotella, F. J., Jorgensen, J. D., Williams, J. M.: Inorg. Chem. *20,* 1802 (1981)
43. Daniels, W., Yersin, H., v. Philipsborn, H., Gliemann, G.: Solid State Commun. *30,* 353 (1979)
44. de Rassenfosse, A., Brasseur, H.: Bull. Soc. Roy. Sci. Liège *4,* 227 (1935)

45. Brasseur, H., de Rassenfosse, A.: ibid. *4,* 281 (1935)
46. Bozorth, R. M., Haworth, F. E.: Phys. Rev. *29,* 223 (1927)
47. Dideberg, O.: private communication, University of Liège (1974)
48. Schindler, J. W., Fukuda, R. C., Adamson, A. W.: J. Am. Chem. Soc. *104,* 3596 (1982)
49. Gliemann, G., Lechner, A.: to be published
50. Ballhausen, C. J.: Introduction to ligand field theory, London – New York, McGraw-Hill 1962
51. Schläfer, H. L., Gliemann, G.: Basic principles of ligand field theory, London, Wiley-Interscience 1969
52. Schwarz, K.: private communication, University of Wien 1975
53. Gray, H. B., Ballhausen, C. J.: J. Am. Chem. Soc. *85,* 260 (1963)
54. Ryskin, A. I., Tkachuk, A. M., Tolstoi, N. A.: Optics and Spectroscopy USSR *17,* 304 (1964)
55. Gray, H. B.: Transition Metal Chemistry *1,* 240 (1965)
56. Moncuit, C.: J. Chim. Phys. *64,* 494 (1967)
57. Mason, W. R., Gray, H. B.: J. Am. Chem. Soc. *90,* 5721 (1968)
58. Piepho, S. B., Schatz, P. N., McCaffery, A. J.: ibid. *91,* 5994 (1969)
59. Interrante, L. V., Messmer, R. P.: Chem. Phys. Lett. *26,* 225 (1974)
60. Interrante, L. V., Messmer, R. P.: in: Ref. 6, p. 382
61. Interrante, L. V.: General Electric Company, private communication 1976
62. Isci, H., Mason, W. R.: Inorg. Chem. *14,* 905 (1975)
63. Viswanath, A. K., Krogh-Jespersen, M. B., Vetuskey, J., Baker, C., Ellenson, W. D., Patterson, H. H.: Mol. Phys. *42,* 1431 (1981)
64. Pereira Neto, J. R., Watari, K., Leite, J. R.: Solid State Commun. *50,* 473 (1984)
65. Marsh, D. G., Miller, J. S.: Inorg. Chem. *15,* 720 (1976)
66. Miller, J. S., Marsh, D. G.: ibid. *15,* 2293 (1976)
67. Cowman, C. D., Gray, H. B.: ibid. *15,* 2823 (1976)
68. Yersin, H.: Habilitationsschrift, University of Regensburg 1980
69. Slater, J. C.: in: Advances in quantum chemistry (ed. Löwdin, P.-O.), Vol. 6, p. 1, New York, Academic Press 1972
70. Johnson, K. H.: in: Advances in quantum chemistry (ed. Löwdin, P.-O.), Vol. 7, p. 143, New York, Academic Press 1973
71. Schwarz, K.: Theor. Chim. Acta *34,* 225 (1974)
72. Schwarz, K.: Phys. Rev. *B5,* 2466 (1972)
73. Rösch, N.: in: Electrons in Finite and Infinite Structures (ed. Phariseau, P., Scheire, L.), p. 1, New York, Plenum Press 1977
74. Lopez, J. P., Yang, C. Y., Case, C. A.: Chem. Phys. Lett. *91,* 353 (1982)
75. Hellwege, K.-H.: Einführung in die Festkörperphysik, Berlin, Springer-Verlag 1976
76. Ziman, J. M.: Principles of the Theory of Solids, London, Cambridge Univ. Press 1972
77. Ashcroft, N. W., Mermin, N. D.: Solid State Physics, New York, Holt Saunders Internat. Editions, 1976
78. Whitmore, D. M.: Phys. Lett. *50A*, 55 (1974)
79. Fritsche, L., Rafat-Mehr, M.: in Ref. 7, p. 97
80. Abarbanel, A. M.: Annals of Physics *91,* 356 (1975)
81. Messmer, R. P., Salahub, D. R.: Phys. Rev. Lett. *35,* 533 (1975)
82. Yersin, H., Gliemann, G., Rössler, U.: Solid State Commun. *21,* 915 (1977)
83. Bullett, D. W.: ibid. *27,* 467 (1978)
84. Whangbo, M.-H., Hoffmann, R.: J. Am. Chem. Soc. *100,* 6093 (1978)
85. Tuszynski, W., Gliemann, G.: Ber. Bunsenges. Phys. Chem., in press
86. Tuszynski, W.: Thesis, University of Regensburg 1976
87. Cusachs, L. C.: J. Chem. Phys. *43* Suppl., 157 (1965)
88. Carroll, D. G., McGlynn, S. P.: J. Chem. Phys. *45,* 3827 (1966)
89. Rousseau, D. L., Butler, M. A., Guggenheim, H. J., Weisman, R. B., Bloch, A. N.: Phys. Rev. B *10,* 2281 (1974)
90. Holzapfel, W., Yersin, H., Gliemann, G.: J. Chem. Phys. *74,* 2124 (1981)
91. Yersin, H., Holzapfel, W., Gliemann, G.: Chemica Scripta *17,* 191 (1981)
92. Schultheiß, R., Hidvegi, I., Gliemann, G.: J. Chem. Phys. *79,* 4167 (1983)
93. v. Ammon, W., Gliemann, G.: ibid. *77,* 2266 (1982)
94. Hidvegi, I., v. Ammon, W., Gliemann, G.: ibid. *76,* 4361 (1982)

95. v. Ammon, W., Hidvegi, I., Gliemann, G.: ibid. *80,* 2837 (1984)
96. Eichhorn, M., Willig, F., Yersin, H.: Chem. Phys. Lett. *81,* 371 (1981)
97. Kuper, C. G., Whitfield, G. D. (eds.): Polarons and Excitons, Edinburg, Oliver and Boyd 1963
98. Haken, H.: Fortschritte der Physik VI, 271, 1958
99. Toyozawa, Y.: in: Vacuum Ultraviolet Radiation Physics (eds. Koch, E.-E., Haensel, R., Kunz, C.), p. 317, Pergamon-Vieweg 1974
100. Reineker, P., Haken, H., Wolf, H. C. (eds.): Organic Molecular Aggregates, Series in Solid State Sciences 49, Berlin, Springer-Verlag 1983
101. Emin, D., Holstein, T.: Phys. Rev. Lett. *36,* 323 (1976)
102. Fugol, I. Ya.: Adv. in Physics *27,* 1 (1978)
103. Zimmerer, G.: J. Luminescence *18/19,* 875 (1979)
104. Williams, R. T., Kabler, M. N., Hayes, W., Stott, J. P.: Phys. Rev. *B14,* 725 (1976)
105. Lushchik, Ch. B.: in: Excitons (ed. Rashba, E. I., Sturge, M. D.), North Holland, p. 505, 1982
106. Rice, S. F., Gray, H. B.: J. Am. Chem. Soc. *105,* 4571 (1983)
107. Bär, L., Gliemann, G.: Chem. Phys. Lett. *108,* 14 (1984)
108. Pertzsch, B., Rössler, U.: Phys. Stat. Sol. (b) *101,* 197 (1980)
109. Pertzsch, B., Rössler, U.: Solid State Commun. *37,* 931 (1981)
110. Rössler, U., Pertzsch, B., Yersin, H.: J. Luminescence *24/25,* 437 (1981)
111. Rössler, U., Yersin, H.: Phys. Rev. *B26,* 3187 (1982)
112. Toyozawa, Y.: in Ref. 100, p. 90
113. Yersin, H., Gliemann, G.: Ann. N. Y. Acad. Sci. *313,* 539 (1978)
114. Yersin, H., Hidvegi, I., Gliemann, G., Stock, M.: Phys. Rev. *B19,* 177 (1979)
115. Stock, M., Yersin, H.: Solid State Commun. *27,* 1305 (1978)
116. Sebastian, L., Weiser, G.: Phys. Rev. Lett. *46,* 1156 (1981)
117. Moncuit, C.: J. de Physique *25,* 833 (1964)
118. Day, P.: Inorg. Chim. Acta Rev. *3,* 81 (1969)
119. Day, P.: J. Am. Chem. Soc. *97,* 1588 (1975)
120. Holzapfel, W.: Thesis, University of Regensburg 1978
120a. Gliemann, G., Holzapfel, W., Yersin, H.: J. de Physique (France) in press
121. Yersin, H., Gliemann, G.: Z. Naturforsch. *30b,* 183 (1975)
122. Yersin, H., Gliemann, G.: Ber. Bunsenges. Phys. Chem. *79,* 1050 (1975)
123. Moncuit, C.: Comptes Rendus de l'Académie des Sciences *252,* 397 (1961)
124. (Moreau-)Colin, M. L.: Bull. Classe Sci. Acad. Roy. Belg. *49,* 973 (1963)
125. Tkachuk, A. M.: Bull. Acad. Sci. (USSR) Phys. Ser. (Engl. Transl.) *27,* 668 (1963)
126. Ryskin, A. I., Tkachuk, A. M., Tolstoi, N. A.: Izb. Akad. Nauk. USSR Ser. Fiz. (Engl. Transl.) *29,* 512 (1965)
127. Tolstoi, N. A., Trofimov, A. K., Tkachuk, A. M., Tkachuk, N. N.: Bull. Acad. Sci. (USSR) Phys. Ser. (Engl. Transl.) *20,* 529 (1956)
128. Ryskin, A. I., Tkachuk, A. M., Tolstoi, N. A.: Opt. Spectrosc. (USSR) (Engl. Transl.) *21,* 31 (1966)
129. Moncuit, C., Poulet, H.: J. Physique Radium *23,* 353 (1962)
130. Yersin, H., Gliemann, G., Räde, H.-S.: Chem. Phys. Lett. *54,* 111 (1978)
131. Bergse, P., Hansen, P. G., Jacobsen, C. F.: Nucl. Instr. and Methods *17,* 325 (1962)
132. Dillinger, R.: Diplomarbeit, University of Regensburg 1982
133. Stock, M., Yersin, H.: Chem. Phys. Lett. *40,* 423 (1976)
134. Genard, J., de Rassenfosse, A.: Bull. Soc. Roy. Sci. Liège *10,* 264 (1941)
135. Musselman, R. L.: Thesis, New Mexico State Univ., Las Cruces 1971
136. Sebastian, L., Weiser, G., Yersin, H.: unpublished results
137. Day, P., Ferguson, J.: J. Chem. Soc., Faraday Trans. 2, *77,* 1579 (1981)
138. Anex, B. G., Musselman, R. L.: J. Phys. Chem. *84,* 883 (1980)
139. Hara, Y.: Chemistry Letters, 1063 (1975)
140. Holzapfel, W., Yersin, H., Gliemann, G., Otto, H. H.: Chem. Phys. Lett. *60,* 304 (1979)
141. Yamada, S.: Bull. Chem. Soc. Jpn. *24,* 125 (1951)
142. v. Ammon, W., Yersin, H., Gliemann, G.: Il Nuovo Cimento *63B,* 3 (1981)
143. Yersin, H., Stock, M.: J. Chem. Phys. *76,* 2136 (1982)

144. Daniels, W., Yersin, H., Stock, M., Gliemann, G.: Solid State Commun. *40,* 937 (1981)
145. Davydov, A. S.: Soviet Phys. Uspekhi (Engl. Transl.) *7,* 145 (1964)
146. McClure, D. S.: Solid State Phys. *8,* 1 (1959)
147. Wolf, H. C.: ibid. *9,* 1 (1959)
148. Craig, D. P., Walmsley, S. H.: Excitons in Molecular Crystals, New York, W. A. Benjamin Inc. 1968
149. Day, P.: in Ref. 6, p. 234
150. Day, P.: J. Mol. Structure *59,* 109 (1980)
151. Hidvegi, I.: Thesis, University of Regensburg 1982
152. Simon, J.: Bull. de la Société Royale des Sciences de Liège *37,* 341 (1968)
153. Kubas, G. J., Jones, J. H.: Inorg. Chem. *13,* 2816 (1974)
154. Gerhardt, V., Pfab, W., Reisinger, J., Yersin, H.: J. Luminescence *18/19,* 357 (1979)
155. Clark, S., Day, P., Huddart, D. J., Ironside, C. N.: J. Chem. Soc., Faraday Trans. 2, *79,* 65 (1983)
156. v. Ammon, W.: Thesis, University of Regensburg 1981
157. Viswanath, A. K., Vetuskey, J., Leighton, R., Krogh-Jespersen, M.-B., Patterson, H. H.: Mol. Phys. *48,* 567 (1983)
158. Yersin, H., Gliemann, G.: unpublished results
159. Tolstoi, N. A., Tkachuk, A. M.: Opt. Spectrosc. (USSR) (Engl. Transl.) *21,* 310 (1966)
160. Daniels, W.: Diplomarbeit, University of Regensburg 1977
161. Sturm, W., Drosdziok, S., Happ, H.: Solid State Commun. *16,* 485 (1975)
162. Freund, A., Roth, S., Ranvaud, R.: J. Appl. Cryst. *7,* 631 (1974)
163. Orbach, R.: in: Optical Properties of Ions in Solids (ed. Di Bartolo), p. 355, New York, Plenum Press 1975
164. Hara, Y., Shirotani, I., Ohashi, Y., Asaumi, K., Minomura, S.: Bull. Chem. Soc. Jpn. *48,* 403 (1975)
165. Ahsbahs, H., Sowa, H., Gliemann, G.: unpublished results
166. Hara, Y., Shirotani, I., Minomura, S.: Chem. Lett. Jpn. 579 (1973)
167. Noack, R. A., Holzapfel, W. B.: in: High-Pressure Science and Technology (ed. Timmerhaus, K. D., Barber, M. S.), p. 748, New York, Plenum Press 1979
168. Leyrer, E.: Diplomarbeit, University of Regensburg 1982
169. Yersin, H., v. Ammon, W., Stock, M., Gliemann, G.: J. Luminescence *18/19,* 774 (1979)
170. Stock, M.: Thesis, University of Regensburg 1977
171. Hidvegi, I.: Diplomarbeit, University of Regensburg 1977
172. Interrante, L. V., Bundy, F. P.: Solid State Commun. *11,* 1641 (1972)
173. Debray, D., Decker, D. L., Saint-James, R.: J. Phys. Lett. (France) *40,* L-127 (1979)
174. Lechner, A., Gliemann, G.: unpublished results
175. Kittel, Ch.: in: Einführung in die Festkörperphysik, p. 266, München, R. Oldenbourg Verlag 1976
176. Yersin, H.: Ber. Bunsenges. Physik. Chemie *80,* 1237 (1976)
177. Dexter, D. L.: J. Chem. Phys. *21,* 836 (1953)
178. Zink, J. I.: Acc. Chem. Res. *11,* 289 (1978)
179. Zink, J. I.: Naturwissenschaften *68,* 507 (1981)
180. Leyrer, E., Zimmermann, F., Zink, J. I., Gliemann, G.: Inorg. Chem. *24,* 102 (1985)
181. Gliemann, G., Hidvegi, I.: to be published

Author-Index Volumes 1–62

GPSR Compliance

The European Union's (EU) General Product Safety Regulation (GPSR) is a set of rules that requires consumer products to be safe and our obligations to ensure this.

If you have any concerns about our products, you can contact us on ProductSafety@springernature.com

In case Publisher is established outside the EU, the EU authorized representative is:

Springer Nature Customer Service Center GmbH
Europaplatz 3
69115 Heidelberg, Germany

Batch number: 10620881

Printed by Printforce, the Netherlands